ECOSOVEREIGNTY

In this book, Omar Dahbour develops the idea of ecosystem sovereignty, calling for a reinterpretation of some essential concepts in political philosophy, including territoriality, self-determination, peoplehood, and sovereignty, in order to make the case for peoples' rights to protect and maintain their natural environments. In doing so, he theorizes current and historical struggles against resource extractions and land grabs, especially by food sovereignty and indigenous rights movements.

The basic idea of ecosovereignty is that peoples living in relation to particular ecosystems have a collective right to ultimate authority over those systems and the resources they contain—provided they manage them sustainably. Dahbour argues that this authority has a legitimacy that overrides that of larger states, at least with regard to matters of environmental management. Ecosovereignty claims may strengthen challenges by peoples to states and corporations seeking to control and transform lands and waters for development, against the wishes of their inhabitants. Dahbour hopes the idea will provide a powerful tool for halting extractivism and ecocide, along with the extreme violence that these processes use against farming, indigenous peoples, and nature.

Connecting political and environmental philosophy in an innovative way, *Ecosovereignty: A Political Principle for the Environmental Crisis* will keep scholars and students informed about an increasingly important topic.

Omar Dahbour is Professor of Philosophy at Hunter College and Graduate School, City University of New York, and author of *Illusion of the Peoples* (2003), *Self-Determination without Nationalism* (2013), and other writings on global ethics, environmental philosophy, and critical theory.

ECOSOVEREIGNTY

A Political Principle for the Environmental Crisis

Omar Dahbour

Routledge
Taylor & Francis Group
NEW YORK AND LONDON

Designed cover image: Getty Images

First published 2025
by Routledge
605 Third Avenue, New York, NY 10158

and by Routledge
4 Park Square, Milton Park, Abingdon, Oxon, OX14 4RN

Routledge is an imprint of the Taylor & Francis Group, an informa business

Library of Congress Cataloging-in-Publication Data
Names: Dahbour, Omar, author.
Title: Ecosovereignty : a political principle for the environmental crisis / Omar Dahbour.
Description: New York, NY : Routledge, 2025. | Includes bibliographical references and index.
Identifiers: LCCN 2024033834 (print) | LCCN 2024033835 (ebook) | ISBN 9781032939025 (hardback) | ISBN 9781032939018 (paperback) | ISBN 9781003568155 (ebook)
Subjects: LCSH: Global environmental change—Political aspects. | Environmental policy. | Sovereignty. | Self-determination, National.
Classification: LCC GE149 .D35 2025 (print) | LCC GE149 (ebook) | DDC 320.601—dc23/eng/20241018
LC record available at https://lccn.loc.gov/2024033834
LC ebook record available at https://lccn.loc.gov/2024033835

ISBN: 978-1-032-93902-5 (hbk)
ISBN: 978-1-032-93901-8 (pbk)
ISBN: 978-1-003-56815-5 (ebk)

DOI: 10.4324/9781003568155

Typeset in Times New Roman
by Apex CoVantage, LLC

"The ecology movement reveals itself in the last analysis as a political . . . movement of liberation. It is political because it confronts the concerted power of big capital, whose vital interests the movement threatens."

Herbert Marcuse, "Ecology and the Critique of Modern Society"

"Human dignity needs a . . . guarantee which can be found only in a new political principle . . . whose validity . . . must comprehend the whole of humanity while its power must remain strictly limited, rooted in and controlled by newly defined territorial entities."

Hannah Arendt, *The Origins of Totalitarianism*

CONTENTS

ACKNOWLEDGMENTS

This book has been long in the making. I want to thank my wife, Catherine Monk, who has supported me throughout this endeavor. Many colleagues, only a few of whom I will mention here, have provided continual support and encouragement, as well as valuable criticisms. Among others, I wish to thank John Agnew, Linda Alcoff, Tom Christiano, Ashley Dawson, Kyle Ferguson, Susan Ginsberg, Carol Gould, Jack Hammond, Ashish Kothari, Mike Menser, Christopher Morris, Fred Neuhouser, Graham Parsons, Carol Rovane, and Sally Sedgwick. I also want to thank my student, Vinh Ly, for valuable research assistance in the writing of this book. Finally, I want to mention a remarkable group of younger scholars, some of them my former students, who have been members of the CUNY Environmental Philosophy Reading Group for the past five years. They have provided me with an intellectual home and been my companions in exploring the rapidly expanding universe of environmental philosophy. Among others in this group, I want to thank Susan Erck, Jon Kwan, Tim Mahoney, Natalie Mathes, Pedro Monque, Tyler Olsen, Manny Rodeiro, Ela Tokar, and Nicole Whalen for their friendship and support.

INTRODUCTION

Violence, Extractivism, and Ecosovereignty

In 1990, Michel Serres published his book, *Le contrat naturel* ("The Natural Contract"), in which he coined the term "war on the world" (Serres 1995). By this, he meant the war that humans were waging against other living species and the environments in which they lived. Because the violence of the attacks seemed to be aimed especially at nonhumans, the term "war" was generally taken to be metaphorical, rather than literal. The "war" constituted the transfer of violent means and actions from the intrahuman conflicts of the early twentieth century to the destruction of nonhuman life in the late twentieth century: from the "war *of* the world" to the "war *on* the world" (Ferguson 2006).

But it was not a metaphor. According to the organization, Global Witness, almost two thousand environmental activists have been murdered in the past ten years alone; a more comprehensive account of the violence remains to be done (Global Witness 2022). These acts have been aimed at persons (and ultimately, peoples) who have attempted to halt the activities of extractive industries around the world that are despoiling natural—and human—environments. Such resistance to extractivism has been met with a response that includes unprecedented levels of violence, including assassinations, torture, massacres, and imprisonment. This violence has *not* been episodic; it has been systematic in its sweep, institutionalized in its depth, and transcontinental in its scale.

Communities across North and South America, Africa, and West and South Asia have been targeted by terror campaigns, either perpetrated by transnational corporations with their own private armies or by state militaries aiding the extractive operations. Hundreds of environmental activists in these regions have been killed by death squads, while countless others imprisoned. Ecosystems have been targeted for the systematic eradication of fish and animal species, and

DOI: 10.4324/9781003568155-1

of native and old-growth trees and other plants (including intact marine environments), which stand in the way of using these lands and waters for industrialized agriculture and forestry.

Such acts take place largely out of sight of educated elites in the industrialized countries, who may be concerned about climate change and species extinctions, yet are oblivious to the daily violence exerted on peoples seeking to live in still intact ecosystems. The principal cause of these actions is the desire of industrial enterprises, and the state governments that abet their activities, to unrestricted access to the remaining natural environments that contain significant biomass and other goods and services that can serve as cheap "raw materials." The eventual destination of these materials, rendered into an array of consumer products, can be traced to the same industrialized populations that tend to ignore the human and nonhuman costs of their consumption. According to a recent summary, "mining and extractive industries, logging and agribusiness were the most common drivers" of the violence (Greenfield 2022). These businesses employed or acted in concert with "hitmen, organized crime groups, and their own governments" to threaten, and ultimately murder, scores of opponents.

Attacks on political leaders and activists are only the most visible form of violence visited upon peoples living in ecosystems deemed valuable for industrial uses. The "slow violence" of transforming environments to make them inhospitable to their native and traditional inhabitants (nonhuman and human) has been going on for much longer, and is far advanced in many parts of the globe (Nixon 2011). Peoples affected—and they number in the millions—have lost their homes, their livelihoods, and sometimes their lives, as they struggle to survive in the face of the industrialization of their world. Of course, also affected are the nonhuman animal and plant species, radically diminished in numbers and increasingly driven to extinction, that suffer from the loss of their habitats.

Some may doubt that such extreme levels of conflict and violence are linked in any necessary way to resource extraction, development programs, or other manifestations of the penetration of ecosystems by industrial methods of agriculture, mining, and manufacturing. But, as mentioned briefly earlier, such actions are part of a comprehensive phenomenon of transforming natural environments into "natural resources" for use in industrial processes, and for capital accumulation and profit. This process, designated as "extractivism" by social scientists, is increasingly driving investment opportunities and financial speculation worldwide (Willow 2019). But it is not a new thing; extractivism has been going on for well over a century, depending on the natural goods involved.

The violence and social conflict are produced by a central fact about most extractivist enterprises: they require the displacement, and at times the decimation, of peoples living within the ecosystems targeted for extractions or "development." Resource extractions, in other words, do not only affect intact ecosystems harboring many plant and animal species. But they also crucially affect the ability of peoples to maintain their livelihoods and their cultures within such ecosystems. This has led some scholars to name the process of extractivism, in part, a "genocide-ecocide nexus," produced by "extractive development" that has become an existential threat, not only to intact natural environments and living beings of all sorts but also to increasing numbers of peoples and cultures resident in them (Dunlap 2021). Such peoples, numbering in the hundreds of millions, live throughout the postcolonial world, in indigenous communities, and in surviving rural societies in industrialized countries (Bennholdt-Thomsen and Mies 1999).

How to respond to this unprecedented level of violence against nature and humanity is fundamentally a question of how to, first, rightly claim and second, effectively gain the political power necessary to bring it to a halt. Philosophy can only really address the former matter, but it is the rightful claims of peoples who seek to protect and enrich ecosystems that establish the basis for the political power that these peoples must eventually acquire. Only in this way can the cruel campaigns for eradicating the last intact elements of nature, within which humans have lived sustainably for many centuries, be stopped.

This book proposes a political principle that would give peoples living within intact ecosystems, or having some essential connection to them, a right of ultimate authority (i.e., sovereignty) over their habitats, and the environmental goods and services contained within them. It is a principle that is designed to delegitimate extractivism, inasmuch as it utilizes ecosystems, not for peoples' livelihoods, but for economic growth and profit. Such uses would be regarded as strictly subordinate to the maintenance of ecosystems for all their living inhabitants, human and nonhuman. The ecosovereignty principle is therefore, first and foremost, a principle of *environmental governance*, establishing the legitimacy of certain governing agencies and governance institutions.

It is important to understand that ecosovereignty is also based on a conception of *environmental justice* that regards extractivism *itself* as an injustice that establishes fundamental inequities between peoples with very different relations to natural environments. While most conceptions of environmental justice focus on the unequal distribution of environmental harms that may eventually result from resource extractions and other industrial uses of natural goods, this book takes a different approach to the problem (Bullard 2018). Following exponents of the "environmentalism of the poor," the primary feature of environmental injustice is found in inequitable resolutions of "ecological distribution conflicts," disputes over ownership, occupancy, and use of natural environments themselves (Guha and Martinez-Alier 1997). Rather than focusing on the "downstream" effects of resource use for consumers in industrial societies, this idea of environmental justice looks to the equitable distribution of (and rights to) intact ecosystems, so that environmental harms that may result from their exploitation can be prevented. In other words, not the discriminatory distribution of harms resulting from industrial processes fueled by resource extractions, but the *extractions themselves* are what is regarded as unjust.

This process of environmental injustice is but one part of a comprehensive ongoing environmental crisis that will occupy the remainder of this century and beyond. But it is crucially important to understand that the crisis is not just "environmental," but it is a crisis of *industrial capitalist society* as a whole. In this larger dimension, the dilemmas of the current moment cannot be understood apart from the ideas of the *critical theory of society* that underlie this study. The first epigraph in this book, from one of Herbert Marcuse's last writings, is one indication of this. Why would the ecology or environmental movement necessarily threaten the interests of corporate capitalism? In an era, when "green capitalism," the "Green New Deal," and other neologisms have become the common coin of environmentalism, this view might seem quaint, even wrong-headed. Is not business, as much as government, a necessary partner in the eradication of "dirty" industrial production processes and antiquated consumer goods, among other potential avenues of cooperation?

This sanguine view cannot survive a closer examination of the bases of environmental destruction in general, and extractivism in particular. This examination must utilize, I contend, one of the most important contributions of critical theory to philosophy—the concept of the *domination of nature.* It is only this concept, developed initially by Max Horkheimer and Theodor Adorno in the 1930s, and later expanded upon by them and other critical theorists, including Marcuse, that can show how the environmental crisis involves more than the exploitation of "external" nature—the ecosystems that provide the natural and material context for human life (Horkheimer and Adorno 2002; Dahbour 2024). It also, crucially, includes the subordination of "inner" nature—the characteristics of human consciousness and behavior that are unneeded and perhaps counterproductive for the role of human beings as units of "abstract labor power" in industrial capitalist economies. From both "natures," industrial capitalism seeks to extract crucial value for accumulation and profit—"cheap nature" from the nonhuman world, and labor power from humans themselves (Patel and Moore 2017).

Other concepts of environmental destruction have based their criticisms on notions of the exploitative functions of industrial technology or the dominating role of beliefs in human "supremacy" (Heidegger 1977; Crist 2019). But only the concept of the domination of nature also highlights the human and political dimensions of the problem, through its critique of the domination of human labor and culture in the service of industrial capitalism. This approach alone can explain why the spread of extractivist economies to previously untouched ecosystems has led to the violence and destruction mentioned earlier. It is a result, not only of the elimination of intact ecosystems—ecocide—but also of the domination and exploitation of human populations by corporations and states seeking exploitable resources. Such populations are either to be employed in such exploitation or removed altogether from the "sacrifice zones" of resource extraction.

Resistance to this domination necessarily takes the form of resistance *to capital accumulation* and *industrial development* as such in a way that transcends the conflicts between capital and labor over the proceeds from economic growth within industrial societies. Without the ability to extract resources from natural systems cheaply, capital accumulation through the subsequent exploitation of labor power expended upon natural materials cannot get going. This was Marcuse's central point in asserting that the "ecology movement" threatens the "vital interests" of capital. But what this resistance to industrial capitalism so far lacks is a comprehensive justification for why peoples have rights to, first, *remain in place* and second, *control their own environments* (in those places). This is the purpose of the ecosovereignty principle, which should become apparent as the argument of this book is developed.

Though the twentieth-century critical theorists played a central role in *naming the problem* of environmental crisis in a way that highlighted its social dimensions, they did not succeed in developing specific political ideas for addressing the crisis. For this, we must turn to other sources, in particular, certain *political realist* thinkers of the past century—Hannah Arendt, above all. Ecosovereignty, I contend, is best understood as a "realist" concept that can be used to address the problem of extractivism that is at the core of the current environmental crisis.

The dominant tendency in contemporary political philosophy has been one that sought to derive political concepts from antecedent moral principles. This latter view, expressed by various utilitarian and contractarian philosophers, of whom John Rawls is the best known, has recently been characterized as the theory of "political moralism" (Williams 2005). The deficiencies of this approach have led some philosophers to espouse a "political realist" approach as an alternative way to develop normative principles that address political problems. While the term "realism" has had multiple meanings in political theory, its importance here is as an alternative to the idea that political concepts are essentially applications of moral principles to social problems. Understanding the distinction between these two doctrines, and why the latter, realist, one better justifies the kind of principle advocated here, is important for assessing its significance and legitimacy.

The critique of political moralism was articulated by Bernard Williams and Raymond Geuss in the 1990s and 2000s. Geuss referred to the theory as the "ethics-first" approach to politics. In this, there were, as he wrote, two steps—first, the postulation of a moral principle, and second, the application of the principle to political actions (Geuss 2008, 8). Williams regarded this theory as asserting the irreducibility of a divide between the moral basis of political theory and the interest-oriented basis of political practice. He called this a "Manichean dualism of soul and body," in which there was a persistent problem of how to bridge the gulf between theory and practice (Williams 2005, 12). He described two ways of doing so, neither satisfactory for understanding how political life actually works. On the one hand, there was an "enactment" model, in which "politics is . . . the instrument of the moral" (Williams 2005, 2). On the other hand, in the "structure" model, "morality offers constraints . . . on what politics can rightfully do."

In contemporary political philosophy, both models of politics as "applied morality" coalesce into a distinction between "ideal" and "nonideal" theory. This distinction, as proposed by John Rawls, leads to an understanding of political theory as embodying three different assumptions, according to Geuss (Geuss 2008, 6). First, politics must be understood and judged according to one or more general (moral) principles. Second, these principles apply to individual human (or moral) agents, primarily. Third, the principles are based upon "ethical intuitions" (e.g., about the rightness or wrongness of human behaviors) "that people in our society purportedly share" (Geuss 2008, 7).

Without these assumptions, the ideal/nonideal theory distinction collapses. Moral principles cannot be "applied" to political actions because that is not what morality properly understood is designed to do. While moral principles are most applicable to individual moral agents, and are often based on particular moral psychologies, neither is directly relevant to political life. Geuss characterizes politics as an object of analysis and theorization in several ways, three of which are relevant here. It is worth noting that none of them have to do with postulating ideal principles or examining moral psychology. First, politics deals in "real motivations," understood as embedded within existing political conflicts and movements. Second, political ideas are concerned with actions, not propositions—and therefore make reference to facts about the world, not to the logical or cognitive characteristics of reasoning. Third, political life is always historically constituted; its reference points are historical events, not psychological intuitions; as Geuss notes, there are no *interesting* "eternal questions" in political philosophy (Geuss 2008, 9–15).

The question for a realist view, however, is about what role this leaves for ideas and conceptualization. If ideal theory posits the wrong kind of ideas—moral ones—for theorizing political life and conflicts, what are the right kind? Here Geuss uses the term, "conceptual tool," for what political theory can offer, and is worth quoting at length about this:

> [There are] problematic situation[s] [in which] the problem itself is often not clearly visible as a problem until one has the answer or a possible set of answers. People, that is, can be at a loss what to do or fail to know what they want because they are confused about what is wrong or what the problem precisely is . . . . In such cases, it is sometimes possible that a certain kind of conceptual innovation may help them. Political theory might . . . discharge the function of providing a new thought-instrument or *conceptual tool* to help particular people understand and define, and thus begin to deal with, certain problems.
>
> *(Geuss 2008, 43–44; italics added)*

A kind of theory that political realism offers, then, is the theory for specific (political) actors in particular (historical) situations.

This is a type of theory that is not unknown in the history of political thought, though it has been less prominent in contemporary political theory. It is, for instance, a characteristic of such philosophers as Aristotle, who wrote about the constitutional requirements of ancient Greek city-states, and Hegel, who theorized the requisites for an "ethical life" in modern European societies. It is equally applicable to the theories of Rousseau, who diagnosed the malaise of inhabitants of authoritarian polities, and of Marx, who addressed the specific oppression of workers in emerging industrial economies. Perhaps more relevant to the present moment, however, is the example of Hannah Arendt. Just as Rawls exemplified the "ideal theorist" who attempts to develop a theory of social justice based on antecedent moral principles, so Arendt was probably the paramount political realist of the last century.

Taking on previously unexamined political phenomena (at least by philosophers) such as imperialism, fascism, and racism, she attempted to theorize both the nature of the problems these ideologies named, and possible solutions to them. One such phenomenon was the tremendous increase in stateless and homeless persons resulting from the wars and revolutions of the mid-twentieth century. Arendt's attempt to transform the concept of the "rights of man" into a "conceptual tool" that could confront the dire conditions of persons who had lost everything but their "humanity" is a key instance of political realism. For those who had been reduced to their bare humanity, a right to membership in a political community (from which other rights necessarily followed) was the most important desideratum. She called such a claim a "right to have rights" (Arendt 1951, 296–97).

Yet, Arendt was also skeptical of the ability or willingness of the sovereign territorial states of her day to grant such a right to stateless persons. So she hypothesized a political principle that could entail the creation of "newly defined territorial entities" other than states, which could recognize and protect such a right. The specific principle remained stillborn in her own work, and the new nonstate entities elusive in practice. In the event, the victorious powers of the World Wars employed the idea of human rights for their own purposes, while dispossessed peoples sought ethnonational states by which *they* could exclude or expel others. Both of these developments were clearly contrary to Arendt's own hopes.

Nevertheless, Arendt's notion that political principles could be developed which could claim validity in dealing with problems at a global scale, while embodying these solutions at a local scale, remains a need today. In the current conjuncture, such problems as those of human migration, poverty, and homelessness with which Arendt was most concerned find their proper description as aspects of a worldwide crisis in the relations of humans to their environments.

The new principle needs to be one that can address at least some aspects of this planetary—that is, global *and* local—crisis driven by extractivism, ecocide, and extreme violence that it is our fate in the present moment to confront. In doing so, such a principle should seek to legitimize political movements that are already engaged in facing these problems. This is an entailment of the realist view that theory most importantly speaks to existing problems and felt experiences of deprivation, oppression, and domination, rather than hypothesizing ideal conditions under which such problems and experiences are absent.

There are certain aspects of such worldwide political challenges that a new principle should confront. First, it should deal, not just with matters of rights, distribution, and recognition of classes, groups, and persons *within* political communities. It should also deal with the conditions of "social reproduction" of the communities as a whole. Concretely, this means confronting the problem of scarcities—*absolute* scarcities—that many political theorists had thought were "solved" by the growth economies of the twentieth century (Ophuls 1977).

Furthermore, the scale and boundaries of political communities must also be considered. In the last generation, the subfield of "global ethics" has contained a lively debate about these matters. Cosmopolitans, nationalists, globalists, localists, federalists, and centralists have all had their say about the legitimate scope and shape of sovereign states and other territorial entities (Walzer 1989; Miller 1995; Gould 2004; Caney 2005; Cohen 2012; Dahbour 2013). But this issue must now be rethought in light of the increasing scarcities and other effects of the depletion of resources and destruction of environments. These problems create new difficulties, as well as new opportunities, for re-imagining, as Arendt hoped, the nature of political territories and sovereign communities. Above all, such a re-imagining must seek a principle or principles that do not try to transcend the limitations of geography and history by offering "global" or "planetary" solutions. It should specify the legitimation conditions for "territorial entities" that can effectively face the crises of extractivism and ecocide on the local and regional scales.

This is because it will be important in determining the efficacy of a political principle able to meet the crisis that it is able to legitimate the authority of political communities at the best scale for solving environmental problems. This scale may vary depending on the problems at hand. Anthropogenic climate change or habitat and biodiversity loss may require political actions at scales different from the problems of environmental injustice and dispossession. In terms of the practices of extractivism that eventuate in ecocide, including the eradication of indigenous livelihoods and cultures, the local scale of the ecosystems and cultures that are threatened is undoubtedly an important context for political authority. Without the ability to make decisions about the environments and territories within which they live, many peoples will be unable to meaningfully participate in, not to say determine, the decisions made by states and markets concerning their value and their future (Leff 2021).

This book proposes a principle that, I contend, would meet these criteria and conditions, and give ecosystems—that is, the people that inhabit them, "ecosystem peoples"—a right to control over their habitats and resources. This principle of *ecosovereignty* would grant ultimate authority over ecosystems and their goods and services to their inhabitants or local users and, where necessary, override the claims of private property owners, corporations, and states. As a preliminary definition (the full statement of the principle is given in Chapter 6), the ecosovereignty principle asserts that *peoples living in distinct environments have ultimate authority concerning the maintenance of the natural goods and services in those environments*. This authority would reside in governance institutions at the scale of ecosystems (e.g., common property institutions), as well as in laws and statutes ratified by constitutional states. But this book is not a study in constitutional design. It is intended to justify the *idea* of ecosovereignty, rather than detailing the specific laws and institutions that may embody it, ones that in any case will necessarily vary by jurisdiction.

The argument for the ecosovereignty principle asserts that it is the best response to the problem of extractivism in that it shifts legitimate authority over ecosystems to those peoples who inhabit them and have an interest in conserving the natural goods and services found in them. While consumers, businesses, governments, and landowners may all claim interests in such environments, their rights, if any, are rendered subordinate by the ecosovereignty principle to those of the "ecosystem peoples" resident there (Dasmann 1989).

Ecosovereignty is, first of all, a theorization of existing social movements opposing extractivism in order to maintain their own livelihoods and ways of life in close connection to natural environments. Two cases of such movements are: agrarian peoples who practice subsistence agriculture and indigenous peoples who live on ancestral lands. But they are not the only cases, merely the most politically organized and articulate at this moment. Other peoples in Europe, South America, and elsewhere are fighting "land grabs" by businesses and governments seeking to profit from extractions of soil, water, minerals, and other "natural resources." All such peoples need political rights that can legitimate resistance to resource extractions.

Such a political right must include rights to territory since this is what is so often at stake in struggles over environmental goods and services. Contemporary political philosophy has been the venue for a vibrant debate about the normative basis of rights to land, territory, and property (Dahbour 2019). It is only in the past few years, however, that this debate has begun to include environmental considerations—despite the by now evident fact that such considerations are the heart of the matter. Now the task is to formulate territorial rights so as to make ecosystem maintenance a principal desideratum. Without this, territories may still be subject to rights claims that allow for the fungibility of real property, and the corresponding dangers of resource exploitation.

Despite the importance of territorial rights for fighting extractivism, they are not sufficient to halt state and market attempts to extract value from natural environments. This is because, whatever importance territoriality has for environmental protection, it cannot counteract states' rights when they are based on reasons of public security, collective welfare, or territorial defense. All of these go to the level of sovereignty claims that, because they are claims to ultimate authority, would override territorial rights.

Sovereignty is today also a subject of vigorous debate about its importance and meaning (Santos Campos and Cadilha 2021). The concept is increasingly linked conceptually and practically to the idea of political self-determination. This is a result of the sovereignty claims made by anticolonial liberation movements against the European seaborne empires in the mid-twentieth century. Sovereignty is now increasingly regarded as a necessary means for achieving self-determination, rather than as a general justification for state power (Rawls 1999). Ecosovereignty embodies this understanding by connecting sovereignty with peoples rather than states and providing a justification based on the importance of using sovereign authority to ensure ecological sustainability.

The use of this concept of sovereign authority is meant to generate equitable resolutions of "ecological distribution conflicts" over limited, and increasingly scarce, natural resources. As indicated earlier, the idea of environmental justice in much of the world today concerns not primarily the disposition of hazardous wastes and other environmental harms, but the ownership and control of environmental benefits. Only when such benefits are securely controlled by peoples interested in maintaining them over time, rather than exploiting them for maximum short-term profit, can the dangers of extractivism be avoided. The name sometimes given to this idea of environmental justice is that of "ecological equity"—the equitable distribution of environmental goods and services (Gadgil 2000). The ecosovereignty principle is designed to enable peoples to achieve ecological equity in distributions of natural goods through the assertion of their sovereignty rights over them.

In the end, this will require systems of environmental governance based on the sovereignty of peoples, rather than that of states, many of which are complicit in extractivist practices. The ecosovereignty principle legitimates such substate systems—common property institutions, community land trusts, and indigenous governments—in terms of their rights to control and maintain the natural environments with which they are concerned. In this way, the ecosovereignty principle adapts key lessons of the anticolonial and self-determination struggles of the past century to the political conflicts of the twenty-first century that have a recognizably environmental dimension—in particular, to the attempt to halt extractivism and prevent the mounting resource scarcities affecting many millions of people around the world.

The argument of this book proceeds in four steps, spread across eight chapters. The first step is to understand the nature of ongoing social movements that have taken upon themselves the task of opposing extractivism in their locales and environs. Chapter 1 examines the food sovereignty movement that is the clearest expression of ecosovereignty in the contemporary world. The concept of food sovereignty will be linked with the study of ecological distribution conflicts that is a key part of the new discipline of political ecology. These conflicts are an important way of understanding struggles over resource extractions around the world. Chapter 2 provides a particular interpretation of the indigenous rights movement, as it has been articulated in the political struggles of native populations, as well as by social scientists and international jurists. The interpretation links indigeneity closely with identities rooted in lands and environments within which indigenous peoples have found meaning and sustenance. Indigenous rights are taken to be a key example of place-based rights more generally and provide a case of thinking through what is at stake in such rights claims.

The second step is to understand how attempts to oppose extractivism require rights to land and territory. Chapter 3 describes the importance that is increasingly attributed to land in the current period—a perhaps surprising development, given the assumed disconnection of economic value from land and other material resources in the past hundred years or more. Yet, it will be shown that this new "conflict zone" is the result of mounting scarcities, as well as shifts in food and energy production in the recent past that have rendered land uniquely valuable again. This is what also makes it the target of intensive political struggles by local communities, states, and corporations. Chapter 4 examines the recent debate over territorial rights that has taken on an environmental dimension and proposes a territorial right based on ecological relationships. If land has again taken on an important role in human social life, it is also becoming the object of conflicts over ownership and use. An ecosystem-based right to territory is shown to be a way of sorting out conflicting claims about land rights so that they do not become a means of increased resource extractions.

The third step in the argument for ecosovereignty concerns sovereignty claims directly. These chapters answer the question as to why sovereignty is the right response to extractivism and ecocide. Chapter 5 examines the recent revival of interest in political sovereignty. Especially in light of the politics of globalization and attempts at "global governance," sovereignty has become a means of reasserting local and democratic rights and claims against larger state and regional bodies. Certain problems that the sovereignty concept is presumed to have are shown to be linked to a particular notion of state sovereignty that is no longer current or necessary as an interpretation of the concept. Chapter 6 shows that this renewed interest in, and renovation of, the idea of sovereignty is in accord with self-determination struggles in the twentieth century that have exhibited an interest in making sovereignty claims. The marked convergence between sovereignty and self-determination in these political conflicts is taken to mean that sovereignty itself is now a "liberatory," rather than a statist, idea. This is a reason for reconceiving self-determination in environmental, rather than exclusively political, terms. The full statement of the ecosovereignty principle is given here, along with various conditions for its full realization.

The fourth and final step is to show how ecosovereignty contributes to the political struggle against extractivism and for the sustainable use of environmental goods and services. This contribution is made in two different ways. Chapter 7 shows how ecosovereignty is instrumental to the realization of a certain conception of environmental justice, and how that may play a critical role in halting extractivism. Ecological equity is shown to be the description of an anti-extractivist way of resolving conflicts over natural resources. This, in turn, is to be understood as a conception of "environmental distributive justice"; and ecosovereignty as a means of realizing this conception of justice. The principle itself expresses the legitimation of a certain sort of environmental governance, and this is the subject of Chapter 8. Only those forms of governance that act to protect natural systems are shown to be legitimate. It is the role of the ecosovereignty principle to embody that understanding of legitimacy in various forms of governance, such as common property, land trusts, and indigenous institutions. These are shown not to be in any essential conflict with territorial states, as long as the latter recognize the rights of the former over ecosystems and the benefits that they contain. Given that potential compatibility, the recognition of ecosovereignty claims in "natural constitutions" is proposed as a way of reconciling possible conflicts between rival authorities. A conclusion will examine how the ecosovereignty principle may play a role in leading to just transitions from industrial capitalist societies to more ecologically sustainable ways of life.

The environmental crisis, it is increasingly clear, is above all a political crisis. This is not only a matter of brute power or force. It is also a conceptual problem, in that ideas justifying political authority—such as territorial rights or self-determination—require adaptation to the need for realizing environmental sustainability and justice. It would be a serious mistake to think that it was sufficient to come up with a notion of environmental justice, for instance, without also considering solutions to the problem of legitimate political authority—that is, who has the right to make decisions capable of implementing such solutions. The question of who legitimately has this right to decide is a crucial one, and enabling those who do to claim the power necessary for these changes to be made is critical.

This is the need that the principle of ecosovereignty is designed to meet: it is a conception of *who* legitimately should have this power, and *why* their having it will be the best means of ending extractivism, preventing ecocide, and ensuring a more just and sustainable future. Doing this is consistent with, inspired by, and hopefully useful for those contemporary social movements whose success is essential to facing the environmental crisis. These movements—for instance, those fighting for the rights of indigenous peoples and for the sovereignty of food producers—are on the front lines of the struggle for environmental justice today, however, unacknowledged they may be in some quarters. It is incumbent on philosophers interested in contributing to solutions to the ongoing destruction of natural environments caused by extractivism to take these movements seriously and assist both their self-understanding and public awareness of their goals. This is the task to which this book seeks to contribute.

References

Arendt, Hannah. 1951. *The Origins of Totalitarianism.* New York: Harcourt Brace Jovanovich.

Bennholdt-Thomsen, Veronika, and Maria Mies. 1999. *The Subsistence Perspective: Beyond the Globalised Economy.* London: Zed Books.

Bullard, Robert. 2018. *Dumping Dixie: Race, Class, and Environmental Quality.* 3rd ed. New York: Routledge.

Caney, Simon. 2005. *Justice Beyond Borders: A Global Political Theory.* Oxford: Oxford University Press.

Cohen, Jean. 2012. *Globalization and Sovereignty: Rethinking Legality, Legitimacy, and Constitutionalism.* Cambridge: Cambridge University Press.

Crist, Eileen. 2019. *Abundant Earth: Toward an Ecological Civilization.* Chicago: University of Chicago Press.

Dahbour, Omar. 2013. *Self-Determination Without Nationalism: A Theory of Postnational Sovereignty.* Philadelphia, PA: Temple University Press.

Dahbour, Omar. 2019. "On the Ecological Blindspot in the Territorial Rights Debate." *Territory, Politics, Governance* 7 (2): 217–32.

Dahbour, Omar. 2024. "The Domination of Nature: A Forgotten Theme in Critical Theory?" *Constellations* 31 (3): 368–81.

Dasmann, Raymond. 1989. "Toward a Biosphere Consciousness." In *The Ends of the Earth: Perspectives on Modern Environmental History*, edited by Donald Worster. Cambridge: Cambridge University Press.

Dunlap, Alexander. 2021. "The Politics of Ecocide, Genocide and Megaprojects: Interrogating Natural Resource Extraction, Identity and the Normalization of Erasure." *Journal of Genocide Research* 23 (2): 212–35.

Ferguson, Niall. 2006. *The War of the World: Twentieth-Century Conflict and the Descent of the West.* New York: Penguin Books.

Gadgil, Madhav. 2000. "Ecology and Equity." In *The Use and Abuse of Nature*, edited by Madhav Gadgil and Ramachandra Guha. Delhi: Oxford University Press.

Geuss, Raymond. 2008. *Philosophy and Real Politics.* Princeton, NJ: Princeton University Press.

Global Witness. 2022. "Decade of Defiance." www.globalwitness.org.

Gould, Carol. 2004. *Globalizing Democracy and Human Rights.* Cambridge: Cambridge University Press.

Greenfield, Patrick. 2022. "More Than 1,700 Green Activists Murdered Since 2012." *Guardian*, September 29.

Guha, Ramachandra, and Joan Martinez-Alier. 1997. *Varieties of Environmentalism: Essays North and South.* Delhi: Oxford University Press.

Heidegger, Martin. 1977. "The Question concerning Technology." In *Basic Writings, from Being and Time (1927)* to *The Task of Thinking (1964)*, edited by David Farrell Krell. San Francisco: Harper & Row, Publishers.
Horkheimer, Max, and Theodor Adorno. 2002. *Dialectic of Enlightenment: Philosophical Fragments*, translated by Edmund Jephcott. Stanford, CA: Stanford University Press.
Leff, Enrique. 2021. *Political Ecology: Deconstructing Capital and Territorializing Life.* New York: Palgrave Macmillan.
Miller, David. 1995. *On Nationality*. Oxford: Clarendon Press.
Nixon, Rob. 2011. *Slow Violence and the Environmentalism of the Poor.* Cambridge, MA: Harvard University Press.
Ophuls, William. 1977. *Ecology and the Politics of Scarcity: Prologue to a Political Theory of the Steady State.* San Francisco, CA: W. H. Freeman.
Patel, Raj, and Jason Moore. 2017. *A History of the World in Seven Cheap Things: A Guide to Capitalism, Nature, and the Future of the Planet.* Berkeley: University of California Press.
Rawls, John. 1999. *The Law of Peoples.* Cambridge, MA: Harvard University Press.
Santos Campos, Andre, and Susana Cadilha, eds. 2021. *Sovereignty as Value.* Lanham, MD: Rowman & Littlefield.
Serres, Michel. 1995. *The Natural Contract*, translated by Elizabeth MacArthur and William Paulson. Ann Arbor: University of Michigan Press.
Walzer, Michael. 1989. *Nation and Universe: Tanner Lectures on Human Values.* Oxford: Oxford University Press.
Williams, Bernard. 2005. "Realism and Moralism in Political Theory." In *In the Beginning Was the Deed: Realism and Moralism in Political Argument.* Princeton, NJ: Princeton University Press.
Willow, Anna. 2019. *Understanding Extractivism: Culture and Power in Natural Resource Disputes.* New York: Routledge.

PART I
The Social Movements

1

FOOD SOVEREIGNTY AND ECOLOGICAL DISTRIBUTION CONFLICTS

The Food Sovereignty Movement

It is important to begin our study of a response to the contemporary environmental crisis by considering the political movement that most clearly employs a concept of sovereignty—the food sovereignty movement. As we will see, the reasons for this have to some extent to do with distinguishing their efforts from prior and concurrent attempts to conceptualize the problems of food and hunger using other terminologies. The shortcomings of these attempts, dating from the 1940s until today, motivated a variety of political actors to coalesce around the idea of food sovereignty in the 1990s, as we will discuss shortly. Because of their use of the sovereignty principle, the food sovereignty movement more directly confronts the political dimension of the environmental crisis than most other responses (with the possible exception of indigenous rights struggles, to be discussed in Chapter 2).

In the food sovereignty movement, the concept of sovereignty—a notion of ultimate political authority—is substituted for other ideas that deal with the problem of food through rights-oriented or distributive means. Whatever their advantages, such ideas do not essentially challenge the present structures of power that ensure the continuation of the contemporary "food system." The sovereignty concept alone asserts the necessity for food producers—hunter-gatherers, peasants, farmers, and fishers—to take power over, and reshape in various ways, the food system that they depend upon for subsistence.

Before we can develop a finer-grained contrast between such rival conceptions, the question, "why food?," should be considered. Why does food occupy a central role in a political movement spanning several continents, with many

DOI: 10.4324/9781003568155-3

local variations? Aside from the obvious point that food is a material requirement for human existence, and therefore necessarily a central concern of all human societies, there are other reasons. For one thing, for most of human history, the vast majority of human beings have been involved in the production of food, for themselves and for others. Even today, approximately one-half of the global population is employed in the production of food, whether for subsistence or as commodities (Altieri, Funes-Monzote, and Petersen 2012, 3). For another, it is the case that the character of the planet's natural environments has been more affected by the modern development of the agricultural system than by any other anthropogenic phenomenon (including global warming). As Eileen Crist writes, "*industrial* food production is humanity's most ecologically destructive activity" (Crist 2019, 173, italics added).

Food is an arena of human need and activity that has been the locus of intensive conflict for centuries. In the past hundred years in particular, especially with the explosion of human populations, concerns about widespread hunger and malnutrition have been persistent. This is not the place to review and assess the extensive discussion of the "neo-Malthusian" argument that population, over time, will outrun the availability of food reserves. The key term, of course, is "over time." Since Malthus himself put forward the thesis in the late eighteenth century, it has been repeatedly "refuted," and just as often revived (Cafaro and Crist 2012). Is it a claim about long- or short-term problems? Does it apply to global, or only to regional, food production? Have technological innovations in agriculture indefinitely postponed, if not altogether removed, the threat of shortages? All these questions defy simple answers.

But one thing is certain: the tremendous growth in global population has changed every aspect of the food system. It has provided a motivation to industrialize food production, if only to make tremendous profits from its increased outputs. It has put pressure on subsistence-oriented food production, straining the ability to provide sustainable food stocks for a stable population. It has even changed the nature of food itself, leading to the introduction of newly manufactured seeds, chemical fertilizers, and the agricultural produce that is the result. All this is to say that the food-population relationship, while perhaps not leading inexorably toward disaster, has led to changes that were unanticipated and perhaps undesired. Accommodations to necessity have in turn provoked thought about what desirable outcomes would be, and whether they are possible any longer in a resource-depleted world.

Since the 1940s, a number of new ideas have been developed for articulating the normative basis for claims to food, especially by populations that now lack any direct access to the means of food production. There have been three major concepts employed in the last eighty years or so to do this: the "right to food" (and accompanying variations on this idea), "food security," and "food sovereignty." The origins of the latter concept (and political movement) in the 1990s follow directly from the perceived inadequacies of the earlier ideas.

The idea of a right to food was part of the general movement to codify and institutionalize a human rights regime at the end of World War II. Such a right was expressed in the UN Declaration on Human Rights (in 1948). Despite some modifications to the original idea in international law and diplomacy, it found expression almost thirty years later in the International Covenant on Economic, Social, and Cultural Rights (in 1976), again as a result of UN diplomatic initiatives (Windfuhr and Jonsén 2005, 19). In the intermediate period, beginning in the 1950s, such a right sometimes found expression as "(a right to) freedom from hunger" (Fairbairn 2010, 19–20).

Either way, the idea had two important characteristics. First, the concept of food was undifferentiated—it specified neither the amount or type of food nor how it was produced or distributed. No metrics for determining the nutritional content of available foods, nor the amounts necessary for sustenance and health, were specified in statements of the right. Second, the addressees of the right were states signatory to the legal instruments specifying the right. States were deemed the best agency for remedying food shortages, as well as the institution with the power to remedy violations of the right.

Both characteristics—the lack of specific food requirements, and recourse to state agencies to address shortages—made statements of the right to food (or freedom from hunger) a means of mobilizing food aid in emergency situations—famines, droughts, and other forms of temporary food shortfall. But the right seemed less useful as a way of addressing endemic lacks or losses of adequate food or nutrition. Part of the reason was that states were not generally in the business of provisioning food for their populations as an ongoing matter. Either food was harvested as a result of subsistence farming or hunting/gathering, or it was distributed through market mechanisms disconnected from subsistence agriculture. It was these mechanisms—subsistence farming and agricultural markets—that seemed to require "reform," in order to remedy chronic and continuing shortfalls or needs outside of temporary crisis situations. States were therefore increasingly perceived to be inadequate venues for addressing problems of hunger and malnutrition resulting from *ongoing* deficiencies in food systems.

The concept of food security was developed to deal with the food needs of peoples in non-emergency situations, where the availability, and perhaps cost, of food were problems. The UN's Food and Agriculture Organization (FAO) was instrumental in formulating changing definitions of food security from the 1970s. There were subtle shifts in these definitions, as states became less involved in food provisioning and allocation. In 1975, food security was defined by the FAO as the "*availability* at all times of adequate food supplies of basic foodstuffs to sustain a steady expansion of food consumption and to offset fluctuations in production and prices." By 2001, the FAO was defining food security as "a situation that exists when all people, at all times, have physical, social and economic *access* to sufficient, safe and nutritious food that meets their dietary needs and food preferences for an active and healthy life" (Patel 2009, 664; italics added).

While "availability" implies the existence of a provisioning agency (i.e., states), "access" suggests that persons could find or locate foods, rather than be provided with them.

The justification for this shift was that the availability of food did not necessarily mean that people could get access to that food. For economic reasons, access might be precluded—for example, because of low family incomes or high food prices. The response was to highlight not the production of food, but its distribution. This meant that economic mechanisms for ensuring food access—the removal of trade barriers and the reduction of commodity prices—were key to ensuring food security. These ideas were encapsulated in a new emphasis on the household, rather than the country, as the locus of food security (Fairbairn 2010, 24–26). This emphasis meant that food security was now a matter of ensuring the proper functioning of markets that could distribute food. Households entered markets as distinct consumers, rather than aggregating in a commons of some sort (e.g., countries, regions, towns, or other small-scale communities) for the procurement of food. Finally, food security became a matter of ensuring that consumers were satisfied with what was available (and affordable). The conditions under which food was produced (grown, harvested) did not enter consideration.

This understanding of food security as the provision, through market mechanisms, of large amounts of cheap food for "global" (i.e., foreign) consumers was both a consequence of changes in the political economy of the late twentieth century, and a cause of the development of a food sovereignty movement. While the food security concept did not itself bring about changes in the "food system," it began to seem increasingly inadequate as an antidote to the negative results of these changes. Some even came to regard the food security and food sovereignty concepts as antithetical for this reason (Schanbacher 2010). Above all, concern with food security, as Raj Patel put it, "avoided discussing the social control of the food system" (Patel 2009, 665). This avoidance was particularly unfortunate given the increase in hunger globally, despite the introduction of new agricultural technologies (i.e., the "green revolution") in the same period. By the end of the 1990s, close to a billion people were affected by hunger or malnutrition, in spite of increases in food production during that same period (Wittman, Desmarais, and Wiebe 2010, 1–2).

Such discordances led to the organization of multiple peasant and small-farmer groups, and their coalescence into a global smallholders' alliance, La Via Campesina [LVC], in 1996. This organization developed and highlighted the food sovereignty concept, as a result of "changes imposed by the industrialization of [food] production and the globalization of agricultural trade" (Wittman, Desmarais, and Wiebe 2010, 5). Before consideration is given to the more specific meanings of food sovereignty espoused by LVC and other groups, we should more clearly specify the problems that such groups were facing. They may be understood under three distinct rubrics: [1] the industrialization of agriculture, [2] the globalization of agricultural markets, and [3] the resultant environmental problems.

The industrialization of agriculture in the "neoliberal" period after 1980 affected the ownership, most importantly, of seeds and of land. In the first case, efforts were made, under the World Trade Organization's provisions on intellectual property rights, to patent seeds and other genetic materials, often passed down for generations among farmers, or freely available in biomass-rich (usually tropical) regions (Shiva 2020). In addition, attempts were made to substitute newly engineered seeds (using genetically modified organisms) for traditional seed stocks.

In the second case, "land grabs" in a number of regions were made as a result of several, sometimes connected, processes. The imposition of austerity measures on governments in order for them to maintain loans from global banks resulted in the "removal of subsidies and other forms of support to small farmers in the South"—thus allowing consolidation of land ownership (Bernstein 2014, 1033). The concentration of agribusiness corporations through mergers and acquisitions enabled larger landholders to further expand their holdings. Finally, the expansion of biofuels production, requiring extensive landholdings that take acreage out of food production, resulted in "appropriation of land [in] areas that are . . . valuable common property resources which generate livelihoods for the poor" (Borras, McMichael, and Scoones 2011, 7).

At the same time as increasing pressure was exerted on peasant and smallholding farmers by agribusiness corporations, often assisted by states, the distribution of food was increasingly affected by so-called market forces—in reality, the large food producers (and their political allies). As the Our World Is Not for Sale Coalition (an anti-World Trade Organization alliance) stated in 2001, "The so-called 'world market' of agricultural products does not exist. What exists is . . . an international trade of surpluses of milk, cereals, and meat dumped primarily by the EU, the US, and other [large producers]" (Wittman, Desmarais, and Wiebe 2010, 204). "Dumping" of food at below-cost prices was therefore a major weapon used to disenfranchise small holders from their land and ability to farm. Other "market distortions" made it difficult for small holders to sell food as commodities, even if they wanted to do so. For instance, Northern countries often supported Northern farmers (with price supports, loans, and other means); markets in Southern countries were often forced to open to Northern food products; and the food processing industry pressured producers to lower the prices on inputs for the products that they manufactured (Windfuhr and Jonsén 2005, 7–8).

One further consequence of these changes in food production was the accelerated destruction of natural environments. The effect of the land grabs was often to push small and poor farmers onto less productive lands, which were then farmed more intensively in order to produce enough for subsistence, with corresponding costs in soil degradation and water depletion (Gomiero, Pimentel, and Paoletti 2011, 115). In addition, the use of increased amounts of fossil fuels and toxic chemicals in food production—in order to increase agricultural profits, or to simply subsist—resulted not only in increased costs of production for farmers but also in negative impacts on biodiversity and human health (Bernstein 2014, 1034).

The impacts of these changes in food systems, especially in the South, were felt initially and primarily at the point of production—on small farms and in rural areas where communities of food producers still produced primarily (although not necessarily, exclusively) for subsistence needs. The attempted creation of a "global" food production system was not generally perceived by food consumers in urban and wealthier rural areas as a matter of forceable imposition. The violence and hardship used to destroy subsistence-oriented agriculture were perpetrated by political and military agents in league with, or directly employed by, wealthier farmers and agricultural corporations. This meant that, if opposition was to arise, it would have to come from those peasants, farmers, and farm workers who knew what was occurring, through personal experience and witness. It was out of such opposition that the food sovereignty concept developed.

In defining food sovereignty, scholars have made lists of various characteristics included in discussions about the concept (Windfuhr and Jonsén 2005, 17). But it may be more useful to refer to a couple of definitions given in different documents published by LVC. One is fairly simple and straightforward:

> Food sovereignty is the right of peoples to healthy and culturally appropriate food produced through ecologically sound and sustainable methods, and their right to define their own food and agricultural systems.
>
> *(Patel 2009, 673)*

But another LVC document from the same period gives a more comprehensive, but also more diffuse and even confusing definition:

> Food sovereignty prioritizes local and national economies and markets and empowers peasant and family farmer-driven agriculture, artisanal fishing, pastoralist-led grazing, and food production, distribution, and consumption based on environmental, social, and economic sustainability.
>
> *(Patel 2009, 674)*

This second definition includes a number of considerations and aspects of the food system that the first left out. But it also blurs the distinction between a more politically actionable conception such as that found in the first definition, and other notions about food, its uses, and its constituencies. Two problems, in particular, should be noted here: a comprehensive definition may include too much, therefore diluting the impact of the principle, and it may be too rights-oriented, thereby missing the element of responsibility for the health of the soil and other ecological components of food production.

While food sovereignty is sometimes regarded as a total approach to building a new food system, what distinguished it initially was the focus on food producers, whose needs would take priority in the construction of such a system. This is justified by their absolutely central role in any system that produces food—other than that of industrialized agriculture. While there will be variations in how food is produced in different ecosystems, all will depend on the protection of food producers, and the natural goods they require to successfully produce sufficient food for themselves and others.

Consumption and distribution of food is dependent on how and what food is produced; it cannot determine them, for the reason that consumers always confront a market in which goods have already been distributed from the point of production. Different patterns of distribution and consumption will follow from distinct forms of (food) production. For instance, local and regional "farmers'" markets will be a natural form of distribution for goods produced within a small geographical scale.

But what of the quantity of food and its availability? For rights to food and food security advocates, this is the crucial issue. The idea that food sovereignty necessarily provides or guarantees the same provisions for food consumers as agroindustrial systems cannot be assumed or implied. Claiming that this is so runs the risk of placing too much emphasis on the possibility that food sovereignty can solve all the problems of the industrial food system. Other ideas and policies will inevitably be needed, perhaps consistent with, but not subsumed under, a food sovereignty regime.

Returning to the more basic definition used by LVC, we may discern three essential elements of the idea: the right of people*s* (as opposed to persons or "people"), the notion of sustainable methods (of food production), and the concept of "self-definition," or self-determination, of food systems. While a right to food, or a right to "freedom from hunger," emphasizes the individual or household consumer as the chief claimant to, or recipient of, food provisions and aid, it is the collectivity—peoples, however defined—that plays this role for food sovereignty advocates. While "people" (or persons) apply to the state, or enter the market, to satisfy their food needs, *peoples* self-organize to produce for themselves. Who are these peoples? At one point, an LVC document uses the term, "nations," to define a peoples' right (Patel 2009, 665). However, it is quite likely that "nation" here means no more than groups of people living together within an ecosystem (that includes or consists mainly of farmlands or fisheries), rather than groups literally defined by ascribed ethnic characteristics, as in the classic definition of nations (Dahbour 2013, 18–41).

What is consistent throughout different documents and discussions of food sovereignty is that peoples are defined geographically, especially with reference to land or territorial habitation. For instance, following upon the mention of the right of "nations" in the 1996 LVC document, it is stated, "We have the right to produce our own food *in our own territory*" (Patel 2009, 665). Raj Patel summarizes the view by arguing, "Food sovereignty has its own geographies"; to argue for food sovereignty "is to demand specific arrangements to govern territory and space" (Patel 2009, 668).

Second, specific means of producing food—often designated as "agroecology"—are needed to clearly characterize food sovereignty as a concept. What is agroecology? We will discuss it more fully next when it is compared to industrial agriculture in terms of their relative strengths and weaknesses. But for now, it may be understood as nonindustrial food production, using the best practices of traditional and modern agricultural polycultures (i.e., simultaneous production of multiple foods in concentrated farmlands or fisheries), within a given ecosystem. It has been regarded as requiring "access of peasants [or similar food producers] to land, seeds, water, credit and local markets" (Altieri and Toledo 2011, 588). This access must be a part of any basic understanding of food sovereignty.

The reason is that industrial food production requires the integration of farming peoples into a food system in which they lose their autonomy over the aforementioned factors of production. Industrialization of food production introduces "economies of scale" that require land tenure systems of large-scale management and ownership. These large-scale farms exist in order to intensively grow single crops for maximum yields. Chemical pesticides must be introduced in order to prevent widespread loss of whole crops in the case of infestations. Seed stocks must be centralized, and therefore centrally owned and maintained, in order to utilize them on industrial scale farmlands. Hydraulic systems must also be centralized in order to manage pollution from chemical run-offs, and in order to allow more intensive cultivation. Finally, the credit and distribution system must be centralized in order to procure the greater credit and market access necessary for handling single large crops that are not grown primarily for local needs. Food sovereignty, if it is accompanied by agroecological methods of production and distribution, eliminates the need for much of this, but does have a different set of requirements, having more to do with localized, continual, and renewable access to basic factors of agricultural production—while foregoing the widespread use of chemical inputs, and of the financial and transport needs of intensive cultivation and long-distance market access.

Third, food sovereignty crucially involves the ability of food producers to define, and beyond that, to govern the food system in which they work: "Food sovereignty . . . ensures that the rights to use and manage our lands, territories, waters, seeds, livestock and biodiversity *are in the hands of those of us who produce food*" (Patel 2009, 666). In one sense, this is not equivalent to "democratic" control of food production, since it is not a majoritarian conception—food consumers must inevitably constitute a simple majority of participants in food systems. Rather, the idea is similar to the socialist conception of "workers' control" of (industrial) production, but now shifted to an agrarian context. Producers are accorded greater rights, because of their dependence upon, and involvement with, the production process.

But there is a crucial difference. While sovereigntist claims in this context concern producers, such producers are also "users" of the product (food). That is, they not only produce food but also consume the food that they produce (whether there is a surplus that can feed other people is another matter). A subsistence system, in other words, is in this way a "closed" system in which there is no sharp separation between producers and consumers (who are usually the same persons). Unlike (neo)liberal ideas of governance that rely on market mechanisms and/or private ownership rights, food sovereignty depends on ideas of self-government to maintain productive systems that directly meet peoples' needs.

This means that the food sovereignty idea asserts peasants', farmers', and other producers' rights to self-determination, particularly over the collective control and management of the food system itself (what else it involves will be considered later). The mechanisms for self-determination will vary, and require some consideration, though they need not be part of the explicit statement of a basic principle of food sovereignty. Where food production is still managed through common property institutions, with collective ownership and control of farming fields or fishing grounds, these institutions are the obvious venue for self-government in this sense. But where such institutions have ceased operating, they will need reconstructing, gradually replacing private (or in some instances, governmental) property ownership systems. While historical institutions, rooted in local traditions, have the greatest chance of easily realizing the promise of food sovereignty for a sustainable food system, this is not the only way. More recent innovations, such as community land trusts, may also be used to re-communalize agrarian, and indeed, some urban, environments (Davis 2010).

Yet, food sovereignty has not gained anything approaching general recognition on the part of international agencies, including those responsible for dealing with food needs. Its lack of incorporation in international legal documents concerned with food production is evidence of this. When the rights of peasants have been acknowledged, as in the UN Declaration on the Rights of Peasants and Other People Working in Rural Areas (2018), food sovereignty remains unacknowledged. This is the case even though, as in the latter document, some aspects of the concept, such as the right to preserve seed stocks, are mentioned.

Why is this? Certainly, part of the answer lies in the sensitivity of states around the use of the term, "sovereignty," for anything other than state sovereignty. The early modern doctrine, which justifies only the sovereignty rights of territorial states, remains hegemonic in international law and diplomacy. Beyond this, however, there are several potential problems with the concept that have led many experts on food systems to reject the idea. Three, in particular, will concern us here: the claim that food sovereignty is anachronistic inasmuch as it ignores the transformation of agriculture that has occurred with the dominance of global capitalism, that it is unable to produce amounts of food comparable to that resulting from industrial agriculture, and that it is environmentally destructive in the sense of eliminating wilderness essential for wild animal species to flourish (and therefore indirectly *contributing to* ecocide).

Food Producers—Peasants or Ecosystem Peoples?

Food sovereignty is a strategy for mobilizing food producers in defense of their environments and the resources needed to subsist in these environments. But why is this mobilization necessary (as opposed to other means of protecting ecosystems)? Are not farmers and fishers already integrated into food systems that may need reform of one sort or another, but are not separate from, and independent of, global markets? Food sovereignty constitutes an attempt to valorize and save the portion of global agriculture that still relies on smallholdings in land, intensive farm labor, and local seed stocks, producing food primarily for local markets. As such, it is potentially a means of also protecting ecosystems and preventing extractivist and ecocidal practices and policies.

But this attempt implies that there are still (farming/fishing) peoples who are subsisting and can continue to subsist within the limits of local, ecosystem-based food production. Is this the case? Before proceeding to answer this question, it should be emphasized that food sovereignty assumes the possibility (whether or not this possibility is prevalent at the moment) of local, self-subsisting commons or communities, based on the food production practices of particular peoples. These are peoples who are defined by purposes and activities that are essentially place-oriented. Two ways to characterize such peoples are as "peasants" and as "ecosystem peoples." Some argue, as we will see next, that there are no peasants anymore since virtually all food producers are integrated to some degree into regional and global markets, in a way that locally oriented producers were not, in the past. If there are no more peasants, for this reason, then there are no peoples dependent on one ecosystem either. Yet, no one disputes that subsistence farming still exists, and is in fact quite prevalent and important for meeting food needs.

So what is the controversy about, exactly? Peasants, as historians understand them, have generally existed within a specific type of land-tenure system involving farming populations tied to estates, and often indentured to the estate owners. They have also been distinguished from "farmers" by the purposes of agricultural cultivation—subsistence primarily, or production for the market (Wolf 1969, xiii–xv). While peasant systems still exist in some places, it is clear that land reforms that emancipated peasants from this system by creating a market in agricultural lands have abolished peasantries as formal legal entities in most of the world (Wolf 1982, 317–18). But subsistence production is still widespread today. Farming populations continue to be tied to lands by subsistence needs, rather than by legal indentures (though informal "debt peonage" can sometimes produce similar ties).

As for "ecosystem peoples," geographers have used this term to refer to groups that subsist within specific ecosystems that sustain the group's existence and culture (Dasmann 1989; Gadgil and Guha 2000). While such peoples include hunting-gathering groups, which are not the main focus of food sovereignty movements, they also encompass farming/fishing communities, as long as these communities are primarily producing for their own sustenance, rather than for markets, or long-distance trade.

Ecosystem peoples are populations who primarily "live in place." They are described by Raymond Dasmann as those able to "survive, year after year and often for centuries or longer, on the resources of a single ecosystem or contiguous and related ecosystems . . . . [They] had achieved a sustainable way of life that did not bring about changes in the natural biota deleterious to their continued existence" (Dasmann 1989, 277). Joan Martinez-Alier has further specified such peoples as doing three things. They [1] "occupy relatively little environmental space"; [2] "managed sustainable agroforestal and agricultural systems"; and [3] "make prudent use of carbon sinks and reservoirs" (Martinez-Alier 2002, 13).

One criticism of this view that ecosystem peoples can legitimately claim sovereignty over their environments is that it relies on a conception of peasant agriculture that no longer exists. The reason is that rural class structure in capitalist societies has eradicated the class of peasants and smallholders, for the most part. But what portion of global agriculture still relies on such small-scale, local resources? According to some sources, as much as one-half of the world's population is involved in such activities, growing food that feeds up to 70% of the total global population (Altieri, Funes-Monzote, and Petersen 2012, 3).

Others think these estimates are wildly exaggerated: at most, a quarter of the global population may still be engaged in some form of subsistence agriculture, while a half or more of this population is now resident in cities (though these are not necessarily mutually exclusive categories, given the widespread use of small-scale urban gardens for nutritional supplementation by the urban poor). Food grown by rural smallholding farmers may be feeding half of the world's people at most, with this proportion likely dropping, as population growth outstrips the farmland acreage necessary for the production of adequate food stocks.

Such empirical debates cannot be decided here. More to the point are the definitional and evaluative dimensions of thought about agricultural landholding and food production. Critics of food sovereignty assert that the concept assumes a way of life that no longer exists: the *via campesina*, the "peasants' way." As Henry Bernstein puts it, "There are no peasants in the world of contemporary capitalist globalization" (Bernstein 2014, 1044). The reason is that the pressures of encroaching market economies force farmers to become "micro-capitalists," selling their food in any market that they can, in order to obtain the liquid capital necessary to maintain their farms, and to expand their landholdings. The result is the eradication of the subsistence farm as a viable economic enterprise: we are all necessarily capitalists now: "there are far fewer petty commodity producers able to reproduce themselves primarily . . . from their own farming in the world today than the numbers of 'peasants' claimed by food sovereignty advocates" (Bernstein 2014, 1044).

From Bernstein's perspective, the real issue in the agricultural regions is one of the proletarianization of farm labor: recognizing that rural economies are now thoroughly capitalist is the only way to understand what justice in the agrarian context could mean. But what is that? Socialism for farmers—that is, collective farms? Reducing the specificity of agrarian struggles to a simulacrum of industrial labor relations is unhelpful, if the goal is to achieve an equitable distribution of ecological goods and services. As the exponents of political ecology have argued, it is the equitable resolution of "ecological distribution conflicts" that is the key to both ecological sustainability and environmental justice in an agrarian context.

What this could mean involves, among other things, learning from the remaining structures of noncapitalist agricultural production, elaborated with twenty-first-century agrarian technologies, to achieve sustainability in food production and justice in land distribution. But the beginning of an alternative is to preserve or revive the wisdom of traditional forms of agrarian life that can provide templates for postcapitalist and postindustrial political ecologies. Here is how some "agroecologists" have put it: "The assemblage of traditional systems still existing in many countries of Latin America, Asia, and Africa comprise a globally important ingenious agricultural heritage . . . yet modern agriculture constantly threatens the sustainability of this inheritance" (Altieri, Funes-Monzote, and Petersen 2012, 4).

Aside from the normative issue of what kind of alternative agriculture to advocate—further "modernization," perhaps along collectivist lines, or revitalization of traditional "agroecological" methods—there is the definitional issue. Does the fact that farmers sometimes participate in market-oriented production in order to accumulate some capital for farm equipment or other needs make them "micro-capitalists"? If so, then farmers have, since even before the expansion of global capital markets, been such proto-capitalists. It may be more important to determine whether the goal for such farmers is subsistence first—production for family and neighbors—or profit in order to continually satisfy these needs through capitalist markets. Occasional participation in regional or global (as opposed to local) markets does not make farmers into capitalists. Only if food production is *primarily* oriented toward market distribution and expansion, rather than subsistence, is this an issue for the food sovereignty approach. Otherwise, food sovereignty will continue to be the practice that protects the local, ecosystemic basis for subsistence on the part of producers, and local consumers as well.

Agroecology as the Basis for Food Security

A second problem often raised for food sovereignty is that it undercuts the efficiency of industrial agriculture, and this efficiency is especially needed now, given the continually growing global population and its food needs. This, indeed, is the chief benefit that is often attributed to industrial agriculture—its ability to produce large quantities of food relatively cheaply, thereby ensuring that the minimum food needs of the world's poor are generally met (Bernstein 2014, 1053). The doubt about food sovereignty, and its accompanying methods of agroecology, is whether this ability to meet global food needs will be sacrificed so that farmers can regain control over food production *for themselves*. So skepticism about food sovereignty is, at least partly, doubt about "beliefs that peasants practicing low (external)-input and labor-intensive farming can feed current and projected world population" (Bernstein 2014, 1056–57).

One way that this problem is often posed is in terms of food security: will food sovereignty for producers result in more food *in*security, as a result both of interruptions in the delivery of food to consumers (so that producers can control and direct access to markets) and of reductions in the productivity of agriculture due to less industrialized means of food production? Food sovereigntists have responded that: first, the industrial system of food production itself results in food insecurity, and this will be magnified in coming decades, as the result of soil exhaustion, global warming, and other effects of industrialization come to fruition. Second, a shift to postindustrial food production—specifically, what is now designated as agroecological methods—does not entail reductions in the *total* production of food, though it may result in declines in productivity for specific crops (e.g., corn or beef). In the case of the latter, however, there may be corresponding gains in nutritional value from changes in food production that improve diets and food quality generally.

Food sovereignty advocates argue that strengthening peoples' connections to farmlands, ecosystems, and place-based livelihoods will—at least in the long run—provide a more stable basis for food security than industrialized production and market-based distribution of food, especially if it involves long-range supply chains and trade circuits. Presumed short-term supply problems should therefore be counterbalanced against long-term problems of soil exhaustion, water shortages, chemical pollution, climate vulnerability, and disease spread that are increasingly endemic to the industrial model of agriculture. In addition, the continuing problem of drastic overpopulation—which political elites persist in ignoring or downplaying—will only make it harder for *any* form of food production to satisfy food needs. The interplay between demographic trends and agricultural systems should therefore also be taken into account.

One aspect of this problem is the apparent disconnection between food sovereignty's emphasis on localized food production (i.e., for local markets) and the growing needs of urban food consumers without direct access to food. While it may be the case that traditional farming methods can feed a local rural population, already connected to food production, it appears doubtful to some that the same system could also feed the many millions of food consumers in the cities (Monbiot 2022, 145–46). A further problem concerns the ownership of specific crops and foodstuffs. For instance, this problem has arisen with respect to the ownership of quinoa (Zerbe 2013). There appears to be a dilemma here. On the one hand, if control of the seed stocks is held by farmers themselves, then they can monopolize growth and distribution, restricting the size of crops and indirectly contributing to hunger in other regions. On the other hand, if control of seeds is taken over by corporations who can transplant and scale up production in other regions, then the livelihood of the farmers who developed the crops may be threatened, while food corporations profit.

How do food sovereignty advocates respond to these problems? Before examining the responses, it should be considered to what extent the doubt or skepticism about food production is even answerable. Skepticism of this sort is notoriously hard to allay, because it is a worry about *the future*—and one based on no definite knowledge or facts, but the *possibility* that change will not provide what current arrangements seem able to do. Some of the indeterminate factors include: [1] whether industrial food production will remain constant, decrease, or increase; [2] how climate change will affect food production generally; [3] whether population growth will continue at current rates, or not; [4] the degree to which urbanization will proceed, diminish, or even reverse. All of these things could have major effects on either industrial or agroecological food production, determining the degree to which they are able to deliver food to consumers.

The upshot of these indeterminate factors is so uncertain that there are even those who do not see an inherent problem with food production as such, at least in the near future. For instance, Vaclav Smil writes, "There appear to be no insurmountable *biophysical* reasons why we could not feed humanity in decades to come while at the same time easing the burden that modern agriculture puts on the biosphere" (Smil 2000, xxvii; italics added). Nevertheless, it is clear that the choice of the socioecological system of food production in various world regions is an important one, and could have major effects, not only on food production as such but also on the lives of the producers, as well as the consumers, themselves. Given the necessity of this choice, food sovereignty advocates make two major points in favor of their approach, including the adoption of agroecological methods that would make autonomy and self-sufficiency for food producers more widely available. First, various improvements in the use of agroecology could greatly improve crop and food yields. Second, the costs incurred and harms produced by industrial agriculture will greatly affect its ability to continue to produce yields seen in the recent past.

Considering agroecology as the basis for food sovereignty means, to begin with, that different standards may need to be used for the measurement of outcomes. For instance, measuring the maximum yields of single crops generally favor industrialized, monocrop agriculture. The standard reason is that economies of scale can produce greater yields of specific crops. But even if this is true, what needs measurement is the *total food produced*, not that of single crops. Traditional and modern agroecological farming is generally multicrop production; the classic example is the "three sisters" of native North American farming—beans, squash, and corn, grown together in the same fields. When this is brought into consideration, economies of scale lose their advantage: "small family farms are much more productive than large farms if total output is considered rather than yield from a single crop" (Altieri and Toledo 2011, 594).

This kind of contrast in results indicates that there may be possibilities for greatly increasing yields without resorting to industrial methods, *if* traditional methods are intensified in various ways in order to produce for a larger population. A number of experimental projects attempting this—and in particular adapting agroecology to a larger scale of production—are under way. Results so far seem to be provisional, and inconclusive. Some assert that large improvements in yield are possible; for example, Altieri et al. claim that "agroecological systems are not limited to producing low outputs . . . . Increases in production of 50–100% are fairly common with most alternative production methods" (Altieri, Funes-Monzote, and Petersen 2012, 6). However, George Monbiot, while conceding that, "*Sometimes* the yields from agroecology are higher than from conventional farming," nevertheless concludes that "it's impossible to feed the world on *low-yield* agroecology" (Monbiot 2022, 177; italics added).

The issue then becomes: Is there any reason to think that agroecology could not be scaled up to encompass food production for the current global population? Two problems that need to be considered in thinking about this are: [1] what sort of socioecological system either inhibits or enables greater food production, though still within the parameters of ecological sustainability, and [2] what is the best mix of population size, human vs. mechanized labor, and urban vs. rural habitation for a sustainable food system run on agroecological principles?

It is important to remember the connection between food sovereignty and agroecological practices. If conditions that allow agroecological methods to be used are possible, then food sovereignty becomes a principal consideration in adopting these methods. This is because, first, nonindustrial food production often eschews seed stocks and chemical additives that aid nonperishability and permit long-distance shipments. The added nutrients and freshness of food will require a greater proximity of peoples to lands on which their food is produced. Second, agroecological methods are often substantially more labor-intensive and therefore will require more people living and working in agrarian environments. Their sovereignty over these environments, at least with respect to food production, will make such livelihood changes easier to accomplish.

In terms of socioecological systems more generally, the comparison of different sorts of agricultural productivity is important because it indicates differential motivations within the different systems. For instance, industrial systems may not give attention to the constituents of a balanced and healthy diet, since they are focused on maximum yields of a limited number of crops. If an industrial system is adapted to producing a variety of foods, the yields in monocrop production will inevitably decline. Another variant is the intentions of subsistence farmers to produce maximum yields. It may be that the lower yields of some traditional farming systems are evidence, "When the subsistence farmer succeeds in providing food, there is no pressure to innovate or to enhance yields" (Altieri, Funes-Monzote, and Petersen 2012, 4). This means that scaling up agroecology for more than subsistence needs is possible, but has not often been tried, since the goals of subsistence may be obtained more readily than production for food consumers who are not also food workers.

This leads to the second problem: for whom should subsistence-oriented farmers be producing food, and who should be involved in this production? Here is where the "quinoa dilemma" becomes significant. Do farmers have the right to monopolize the production of specific foods? While denying specific food stuffs to large populations of food consumers helps no one, producers or consumers, taking control of seed stocks out of the hands of growers and placing it into the vagaries of global commodities markets can threaten ruin for the original producers. Measures need to be taken to ensure that the farmers have a steady source of income, and control over the seed stocks necessary for satisfying local consumption. Alliances between farmers in different regions, who can share seeds and information, is now developing through LVC and other organizations. This can help to achieve both sufficiency and equity, as long as the intervention of agricultural corporations is prevented.

To return to another related problem, the population issue can only be avoided for so long. It may be that, in any ecologically sustainable form of food production (perhaps excluding industrial agriculture for this reason), population growth, *beyond a certain point*, will continually threaten to outrun available food supplies. Whether or not this neo-Malthusian scenario materializes, it is obviously advantageous to find ways to end population growth, and eventually to reduce the numbers of food consumers.

In addition, the issue of labor—whether food consumers will also, and to what extent, be food producers—must be considered. A less mechanized and chemically-intensive form of food production will inevitably require more labor inputs. Some studies have shown that increasing labor inputs, for instance in urban food production, can produce significantly higher outputs (McDougall, Kristiansen, and Rader 2019, 131). Is this a good or bad thing? It may be that one of the anticipated *advantages* of prioritizing food producers who grow for themselves primarily is that it will require the greater involvement of large numbers of people in food production. Certainly, industrial economies have shown themselves capable of producing chronic and increasing unemployment, as they move through successive "industrial revolutions." Involving more people in food production could be one response to this employment crisis.

Of course, it will require large-scale changes in the mix of rural-urban habitation, even with the increasing development of urban agriculture. But what is the alternative? In Mike Davis's famous phrase, it may be simply a "planet of slums" (Davis 2005). If industrial agriculture continues, with its extreme separation of city and country, and the continuing eradication of productive farming communities, the result may be that there will be "little room for the hundreds of millions of rural and urban poor in [an] integrated global market. They are confined to giant suburban *favelas* . . . or to rural reservations, where they are . . . increasingly vulnerable to hunger" (Bello and Baviera 2010, 70–71).

In any case, it is only "cheap fossil energy that allows us to provision large urban populations" (Orr 1994, 622). Much of this fossil fuel use, of course, is dedicated to producing food in industrialized farms with high energy consumption. A perhaps unanticipated advantage of food sovereignty is the prospect of redressing the imbalance between rural and urban populations that creates vast numbers of poor food consumers, cut off from food production, *and* stable employment in industrial enterprises. Moving to more labor-intensive farming provides an opportunity to re-employ in agriculture many of these people. Of course, this is not to minimize the difficulties of making such a transition to a more agrarian world, despite its potential advantages, and attractions for the disenfranchised urban poor (Orr 1994, 623).

But the urbanization that has characterized the last century or so is very likely reaching its limit, despite current trends. Today, the "megacities" of the world exist essentially as "entropic black holes" that absorb huge amounts of food and energy, while generating equally extensive wastes, most of which do not get recycled into agriculture, as in agroecological systems (Smaje 2023, 54). Despite a global economy based on fossil fuels, industrial production of food and other necessities, and the attempted creation of consumer markets for long-distance trade, urbanization has absorbed only about half of the world's population (Smaje 2023, 122). Without the "low-cost supply chains" that have very likely reached their limits in terms of affordability, cities will "become responsible for their own food, energy, and industrial outputs" (Smaje 2023, 124). In the coming years, the result may be a "net flow of people from increasingly disordered urban areas to rural areas where the possibility of creating a farmed livelihood exists" (Smaje 2023, 129).

Finally, the doubts about food sovereignty in terms of the productivity of agroecology are also mitigated once the tremendous costs and dangers of the industrial food system are recognized. These costs are fourfold: land grabs that dispossess and impoverish farmers, pressures on food prices that contribute to hunger, built-in waste in food distribution systems, and land degradation that reduces available cropland. The overall problem of relying on industrial food production, what they define as the "food challenge," is summarized by Miguel Altieri and colleagues as a world in which there is "a shrinking arable land base (which is also being diverted to produce biofuels), with less and more expensive petroleum, increasingly limited supplies of water and nitrogen, and . . . a scenario of rapidly changing climate, social unrest, and economic uncertainty" (Altieri, Funes-Monzote, and Petersen 2012, 1–2).

Clearly, the loss of land by millions of farmers through the contemporary equivalent of the early modern enclosure movements in Europe has been a huge driver of urban poverty and hunger. The centralization of land ownership, monocrop cultivation, mechanization of farm labor, and expense of chemical and fuel use in farming have all contributed to the land grabs that resulted in the problems just mentioned. These are, inevitably, impacts of industrial agriculture that have had the paradoxical effect of increasing food production and hunger *at the same time* (Bello and Baviera 2010, 70–71). Furthermore, it is not only the land grabs that have contributed to these results. It is also the increases in food prices, often as a result of the use of fossil fuels, that have caused a reappearance of malnutrition and hunger in the last generation, despite increased efficiencies in food production (Chappell and LaValle 2011, 17).

Beyond this lack of connection between food production and food consumption, it is also the case that industrial agriculture has, despite increased efficiencies in *production*, introduced increasing *in*efficiencies in distribution. This tremendous waste of food, as it travels from points of production to far-flung consumers, is a result of a food system that "relies on regional specialization and long-distance transport." The resultant waste may in some cases reach or exceed 50 percent of the food produced (Wittman 2010, 93). Finally, the effect of the industrial regime on the land itself has been notable: land degradation, in particular, soil erosion as a result of mechanized tillage, among other things, has already caused significant losses in land available for farming, and may, according to one estimate, reduce food production by some 30 percent in the next fifty years (Pimentel and Burgess 2013, 454).

In the end, food security is a bit of a gamble no matter which socioecological system is employed. But two things are certain. Without stabilization, and eventually reduction, of world population, no system will ultimately be able to stand the strain of constantly increasing production, while maintaining material reserves of natural goods (e.g., land, water) sufficient for satisfying needs. There is no "sustainable" development or growth that is possible for food production.

This means that arguing that industrial production, with all its attendant costs and dangers, is the only system that has the possibility of feeding a constantly growing population is a nonstarter. We have already long passed the limits of a sustainable world population, given the coming intensification of crises on multiple fronts.

Furthermore, whatever the future of non-food producing consumer populations, we know that further industrialization of agriculture will lead to immiseration and displacement of the twenty-five to fifty percent of the world's peoples still involved with, and to some degree self-sufficient in, food production. The consequences of this are best *not* imagined. Far better to try a path based on the accumulated knowledge and experience of centuries of traditional food producers, in combination with new agroecological methods for the intensification of food yields, where possible. But to do that, such producers will need some independence from, and protection against, the food corporations and their governmental agents trying to seize lands and waters, and profit from the extraction of maximum amounts of biomass in the shortest possible time. In other words, what is needed is sovereignty for those food producers, and their allies, willing to resist these efforts.

Food Sovereignty and Biodiversity

A final criticism of food sovereignty is that it establishes a presumption that agriculture takes priority over wilderness, and that if there is something like a "right to farm," it will provide a justification for sidelining wilderness preservation even further (Monbiot 2022, 144). Yet, it is not clear either that food sovereignty has such a presumption, that it is any worse than the effects of industrial agriculture in this regard, or that agroecology does not have its own means available for dealing with wilderness loss. In any case, it should be remembered that it is industrial agriculture that has had the most large-scale devastating effects on wilderness, through deforestation, soil erosion, water pollution, and so forth. While food sovereignty is an argument for keeping peoples on (and in control of) land, this does not *necessarily* result in wilderness loss, while it is clear that industrial agriculture has devastated biodiversity in multiple settings. The argument for continuing it is that it might more efficiently produce food on a given acreage than agroecological farming methods. But as we have seen, that is not an eventuality, and may even be completely false, once proper comparative measures are made.

The real issue may be whether food production is aimed at increasing productivity in yields, or at sufficiency in satisfying needs, mainly of local populations (Smaje 2023, 58). In addition, the techniques of "ecological intensification"—"enhancement of ecosystem services to complement or substitute for the role of anthropogenic inputs in maintaining or increasing [agricultural] yields"—also suggest means of *integrating* human cultivation with wilderness preservation. Examples include crop diversification for pest control, animal fertilizer recycling, and the use of plantings to improve soil quality (MacLaren et al. 2022).

Ultimately, the criticisms that have been made of food sovereignty rely on forgetting that industrial agriculture's effects on natural environments have been tremendously destructive. While food sovereignty by itself is not a remedy for all these effects, it does depart markedly from several crucial aspects of industrialization, by attempting to more fully integrate an agrarian human population with the ecosystems on which they ultimately depend—for soil nutrients, for fresh water, and for processing solar energy into natural fertilizers. The substitution of agroecological methods, which food sovereignty is designed to protect and encourage, will have a number of positive impacts toward restoring the health of ecosystems, whatever other problems it may not fully solve.

In addition, it is important to remember that food sovereignty is a solution to a human problem—the destruction of intact, ecosystem-based farming communities, often relatively self-sufficient for food and water needs, and operating within the parameters of a *natural economy*. That it *is* a natural economy and not a political economy organized around industrial production and long-distance trade indicates that the consequences for biodiversity are going to be less, perhaps much less, severe. This is in the first place because agroecological methods that enable food producers to be self-sufficient depend on the maintenance of large amounts of biomass, instead of injections of fossil fuel-based energy inputs and chemical fertilizers. Dependence on a natural economy means that food producers are highly motivated to preserve ecological goods and services such as clean water, healthy soil, diverse animal (especially bird) life for pest control, and varied plantings that can survive adverse climatic conditions. There is also a "self-limiting dynamic" in subsistence agriculture that restricts the effects of the agrarian system on wildlife and native plant species. Subsistence farmers are "driven by the need to feed themselves and their communities, and not to maximize financial returns" (Smaje 2023, 74).

In contrast, industrial agriculture comprises a monocrop food system designed to provide large quantities of a simplified diet to food consumers in urban areas otherwise unable to provide for themselves (Smaje 2023, 73). But to do this, it requires replacing sustainable agrarian civilizations that have persisted for millennia, feeding most of the world's human beings, without destroying (even if they have in certain ways transformed) natural systems. Instead, the industrial system generates "food deserts" (for humans, as well as other species) as a result of monocrop cultivation. It results in soil exhaustion and erosion that reduces living matter across the ecosystem. Water reserves (e.g., lakes, rivers, and aquifers) are polluted and fish stocks are decimated by run-offs from Concentrated Animal Feeding Operations (CAFOs), and other industrial-scale farms and fisheries. The effects of all this and more on biodiversity are devastating. No comparably scaled effects result from subsistence farming and fishing, even in historically concentrated agricultural zones such as the Indo-Gangetic Plain or the Yellow River Basin.

In the end, advocacy of food sovereignty is an *ethical* choice to affirm a certain way of life—self-sustaining, co-dependent with other species, and embedded within ecosystems—over another that relies on the dubious products of industrial economies. To some extent, of course, food sovereignty (or perhaps ecosovereignty more generally) is a practice that *cannot* solve all environmental and social problems (what could?). But such solutions will invariably involve reducing both the industrial production of human needs-satisfactions and the continuing growth of human populations. By re-establishing a sustainable model of food production, however, food sovereignty may contribute to, though it is no substitute for, more ecologically sustainable practices of human fertility, habitation, and energy use.

Food Sovereignty vs. Consumer Sovereignty

Regarding food sovereignty as part of a reaction to the industrialization of agriculture should lead us to consider the larger context of this reaction. This book constitutes an argument for the rights of ecosystem peoples to claim authority over the conditions requisite for their sustainability (and that of the environments that support them). But this is to regard food sovereignty as not only a reaffirmation of a certain sort of agrarian life and politics but also a part of a larger movement that seeks to take control of the conditions of life away from markets and states and reorient them to commons institutions that are not inherently extractivist.

This is the way in which food sovereignty has an ethical dimension; it is not as if opposing food sovereignty was equivalent to a universalist espousal of "food for everyone." Focusing on food security, and rejecting food sovereignty, is equivalent to an ethical position of a different sort—the advocacy of "consumer sovereignty." Organizing a food system on the basis of consumers' needs for food implies that these needs are determinable independently of any particular form of food production. But this is not the case. Needs are determined by the capacity of different systems for satisfying them—there is no simple need for food as such. "Global markets" in food, as indicated earlier, are not truly open ("free") markets, but distribution systems for selling ("dumping") certain foods (e.g., cereals, dairy products, and meat) at low prices in order to obtain greater market share for their (industrial) producers (Wittman, Desmarais, and Wiebe 2010).

Consumer sovereignty, upon which the idea of food security as a principal desideratum is based, is an oxymoron. It is really market sovereignty—and its determination by powerful agribusinesses that can produce for multiple markets—that fixes the extent and limit of food needs in such a system. Food consumers can claim no independence or priority in such a system. Their needs become what the system can satisfy, not what they might prefer in some ideal setting or situation. So who has the power to determine what these needs will be—who has sovereignty in this case—is what is really at stake here. Food sovereignty asserts the importance of *self-determination* by food producers in the design and implementation of a food system; food security/consumer sovereignty asserts the importance of efficiently supplying a market that can provide large quantities of *some* products that consumers will buy. Which ones? Those that are able to be most efficiently produced in large-scale industrial enterprises.

This is not the place to discuss the general relation between sovereignty and self-determination (this will be considered in Chapter 6). But once it is clear that food sovereignty *does* constitute a self-determination claim, the practical implications of the claim bear some consideration, so that what is at stake can become clearer. Two concepts that are correlative with food sovereignty as a self-determination claim are especially important here—those of "common-wealth" and of "livelihood." Both provide a necessary context for, and content to, food sovereignty as an aspect of self-determination for food producers.

Food, and the natural materials that go into producing it, is clearly an important part of the common-wealth that is a feature of commons institutions to maintain. Commons is the name for all collective organizations of users of various ecosystem goods and services—for example, forests, farmlands, rivers, fishing grounds—who act to conserve these goods for themselves and future users. The maintenance of commonwealth in land, water, and other environmental goods is therefore a primary goal of such commons (e.g., peoples who inhabit particular villages, river basins, or pasturage).

In the contemporary world, commons seek to protect their natural wealth from extractivists who would deplete it to satisfy consumer demands elsewhere. These commons organizations therefore seek a kind of self-determination, of which food sovereignty is a specific *functional* equivalent. At least with regard to food and its constituents, peoples living in commons need to claim self-determination, in order to maintain their commonwealth from external threats.

An important question is whether this commonwealth can be maintained at a scale sufficient to feed large numbers of people—consumers as well as producers. This is certainly the aim of food sovereignty and agroecology advocates. As Altieri states,

> Whether the potential and spread of local agroecological innovations . . .[are] scaled up to reach all the small farmers of a region depends on the ability of . . . farmers [to] gain increasing access to agroecological knowledge as well as to land, seeds, government services, solidarity markets, and so on.
>
> *(Altieri, Funes-Monzote, and Petersen 2012, 12)*

So commons in the current era cannot remain traditional village or rural associations only—though they may sometimes start as that. The commons organizations that might claim self-determination, beginning with food sovereignty, will include many different kinds of peoples, though all will be concerned to protect and maintain the natural goods necessary for self-sufficiency in food production.

And not only this. Also necessary are other elements of "livelihood"—the conditions for maintaining independent lives outside of market and state relations. For instance, energy (production) and the built environment (for human habitation) might also be considered important aspects of livelihood. In the era of anthropogenic climate change, the ways in which energy is produced become a crucial component, not only of food production itself but also of the very conditions of life for humans and other species. As Philip McMichael puts it, "Given the conjunction of food, energy, and climate crises, the modernist prejudice against smallholder farming, fishing, and pastoralism needs reversing" (McMichael 2010, 174). It becomes important to take control of energy systems, both for food production and for living in a sustainable way. The "modernist," hyperindustrialized world of energy wastage and overuse for frivolous, economic, or military purposes must be "depowered" through transitions to solar-based energy sources. This, too, would be a benefit of food sovereignty's emphasis on the protection and nourishing of small-scale agriculture.

Human habitation is another aspect of livelihood, in addition to the food and energy aspects. Industrial societies rely on massive investment in artificial infrastructure to provide habitation for its workers and consumers. One recent study estimated the extent of such infrastructure, which consists of buildings for work and living, transport infrastructure of roads, rails, airports, and various vehicles, as well as industrial farming operations, with their monocrop farms and farm machinery, at thirty trillion tons (Purdy 2019, 77–78). Whether or not this artificial infrastructure is necessary for human livelihood is directly connected to the mode of living or form of life that is prevalent. Nonindustrial societies will need much less of this sort of thing because their livelihood is more closely connected to the natural processes of living within ecosystems, rather than in human-constructed environments.

Livelihoods in general consist of the "capabilities, assets (including both material and social resources), and activities for a means of living" (Chambers and Conway 1992, 6). Livelihood sovereignty—as an extension of, and generalization from, food sovereignty—means the authority over these capabilities and assets sufficient to maintain them over time. When peoples have such sovereignty, they are able to conserve the necessary resources to enable them to maintain the livelihood of their communities across generations. However, when a society "continually draws down a livelihood's asset . . . without the replacement of that capital in some form . . . [it] is usually deemed unsustainable" (Carr 2015, 335). Food sovereignty is but one step toward establishing sustainable livelihoods.

The deeper meaning of food (and livelihood) sovereignty is the realization that a revaluation of modern industrial life is required. When applied to agrarian cultures, industrialization has resulted in "unsustainable monocultures of an energy-intensive industrial agriculture" (McMichael 2010, 171). Food, livelihood, and ultimately, ecosystem sovereignty is the assertion of an "alternative modernity" that aims at "restoring agriculture from its historical subordination" to industrialization (McMichael 2010, 173). To do this, it is not enough to protect the increasingly endangered ecosystems that remain relatively intact in the face of the onslaught of industrial agriculture. Efforts must also be made to "*rebuild* small-scale, diverse agricultural systems to resolve modern problems of environmental degradation" (McMichael 2010, 171; italics added). Such efforts potentially will have a number of beneficial effects, not only on ecosystems as such but also for human life in general. Rebuilding sustainable ecosystems, for instance, "offers great potential to generate green jobs and revitalize rural areas" (Gomiero, Pimentel, and Paoletti 2011, 114).

But this brings us back to the sovereignty concept: only if peoples can find the power to assert their difference from industrial-capitalist civilization can they make changes that will result in greater control over their foodways, livelihood assets, and ecosystems. It is the "uniquely constructive power of political sovereignty," which alone will enable peoples to choose "what sort of species we are going to be"—as opposed to being on the receiving end of the power and wealth of states and markets that retain their own sovereignty to make these choices (Purdy 2019, 93–94).

What does claiming this sort of sovereignty—especially over food production—require? In the first place, it will need to be applied to the cause of land redistribution and its corollary, territorial rights. In some instances, land redistribution may mean simply "affirming and protecting pre-existing land access and occupancy by poor peasants whose tenure is 'insecure'" (Borras and Franco 2010, 111). It may also involve a more active transfer of control from states and private landed classes to the "landless and near-landless working poor" (Borras and Franco 2010, 109). One instance of this might be the Landless Workers' Movement in Brazil, which has had some success in redistributing agrarian land to the urban poor (Ondetti 2008; Wolford 2010). All of this amounts to an assertion of self-determination claims focused on land and the materials for producing food that will enable peoples to take back control of their environments, and in particular their ability to feed themselves.

The Food Sovereignty Movement and Ecological Distribution Conflicts

The importance of self-determination points to a central feature of food sovereignty: it is fundamentally a theory of social *conflict*. Of course, the kind of conflict is clearly distinct from other forms concerned with class relations or social identities. But the political element in food sovereignty—the assertion of self-determination claims—means that different types of socioecological systems (for instance, industrial or agrarian), that use natural goods and services in different ways, lead to different political goals to ensure these uses. These goals will be necessarily incompatible with one another and productive of political conflicts and struggles (Le Billon 2015).

The type of conflict involved here is that of "ecological distribution conflicts" (EDCs)—that is, conflicts over the distribution of ecological benefits and burdens. The reason why conflict is inherent between socioecological systems is the tremendous variability in ways that humans have of using natural goods. Such conflicts are the result, as Joan Martinez-Alier has put it, of the fact that "humans have no biological instructions on the exosomatic use of energy and materials" (Martinez-Alier 2002, 70). Consequently, differences between types of use, and conflicts over which should prevail, have been present throughout human history. With the tremendous increase in human populations in the past two hundred years, and the corresponding resource depletion that certain forms of use entail, such conflicts have occurred at an accelerating rate (Gray 2003).

These conflicts involve not only confrontations between distinct cultures rooted in different socioecological systems, but also between social classes over land, water, and other resources. It is this dimension that has led some political ecologists to refer to the subject as the "environmentalism of the poor":

> Many social conflicts today, and in history, have an ecological content, with the poor trying to retain under their control the environmental resources and services they need for livelihood, and which are threatened by state takeover or by the advance of the generalized market system.
>
> *(Martinez-Alier 2002, 263)*

These conflicts, especially (though not exclusively) today in the "Global South," involve struggles by hundreds of millions of people against the encroachments of the state and market. Martinez-Alier lists twenty-three different types of (on-going) ecological distribution conflicts in his 2002 book, most involving such struggles over one or another resource (Martinez-Alier 2002, 258–60; see also Martinez-Alier 2023).

Such conflicts have tended to pit ecosystem peoples against others who seek to extract materials or (economic) value from the environments upon which the former are dependent for their sustenance. Some have characterized this conflict as a struggle between different social classes, who exist on different geographical scales in terms of the resources necessary for their survival. While ecosystem peoples, dependent on particular environments, seek sovereignty over them in order to survive, a distinct class operates at a broader scale than specific ecosystems and exploits the resources of various regions in a nonsustainable way. The key feature of this class has been described as the ability to "continue to survive despite destructive exploitation of local environments . . . . [They] can draw on the total resources of the biosphere to override the controls previously exercised through the productive limits of local animal and plant populations" (Dasmann 1989, 277–78).

They can survive, but not the ecosystems they use (up) for their own consumption, nor the peoples who depend on those ecosystems. At one time, Raymond Dasmann referred to this group as "biosphere people," since they made use of resources from around the globe to meet their own needs (and desires). But he came to reject the term since it implied that such people had "achieved the same sort of symbiosis with the total biosphere that characterizes ecosystem people and their home environments" (Dasmann 1989, 278). Instead, Madhav Gadgil and Ramachandra Guha have coined another term "omnivores" for such people—those who have the power to "avail themselves of the resources of the entire country, and indeed the globe" (Gadgil and Guha 2000, viii). Such omnivores operate through the imposition of economic development policies imposed by an "iron triangle" of state agencies, corporate interests, and the utilization of new technologies designed for maximum exploitation of resources. This combination of state, market, and new technology is the basis for the aggressive policies of extractivism mentioned in the Introduction, and antithetical to sustainable (especially agrarian) socioecological systems around the world.

There are two results of this process: "resource depletion, habitat fragmentation, and species loss," in specific ecosystems; and the dispersion of the peoples dependent on those ecosystems, resulting in the creation of a third class—"ecological refugees." Ecological refugees are sometimes confused with "climate refugees," those fleeing the effects of anthropogenic climate change in flood zones, coastal areas, desert regions, and other places vulnerable to storms, flooding, drought, and fire (Vince 2022). Ecological refugees, by contrast, are peoples displaced by the processes of land enclosures, market pricing fluctuations, debt burdens, deforestation, fishing stock depletion, water pollution of coastal zones, and other effects of market mechanisms and state policies that aid capital accumulation for wealthy investors, landowners, and agricultural corporations, while destroying sustainable ecosystems in which such peoples lived.

To generalize the nature of these sorts of EDCs, they may be put into two broad categories. On the one hand, conflicts have occurred when omnivores seek to "extract"—that is use up without replenishment—"resources" from lands and waters which have been used for sustainable livelihoods by ecosystem peoples. These direct, one-time extractions may be of minerals (e.g., coal), biomass (e.g., forests), or water (e.g., aquifers). Traditional agrarian usage of such lands and waters is usually nonintensive and therefore incapable of supporting an industrializing society on a long-term basis. Nevertheless, the resources may be capable of being extracted as "raw materials" that contribute to industrial production and consumption elsewhere. On the other hand, peoples who have transformed environments for subsistence agriculture may also be in conflict with other producers ("modern" farmers, businesses, or governments), who are attempting to transform such areas for longer-term use by industrial agriculture, usually monocrop production. This also involves the extraction of natural goods (such as soil or groundwater), and, in turn, renders the lands unfit for subsistence use. Both conflicts pit local food producers against global consumers—and the businesses that produce for *them*—who may benefit from the extraction of nonrenewable goods or from agricultural monocultures.

Food sovereignty, finally, is part of this struggle; but it is a peculiarly modern (or "postmodern") effort to valorize what is at risk of being lost—the ability of peoples, rooted in particular places, to sustain themselves within specific natural environments. The claims of these peoples for self-determination—independence and self-sufficiency—is, then, at the core of the struggle for food sovereignty. But there is, in addition, another dimension that often manifests itself as an EDC, sometimes in tandem with the struggles of food producers to survive. This is the fight to preserve indigenous cultures and traditions, many of which have also played a role in sustaining peoples' livelihoods within particular ecosystems. This struggle has also been, in part, about maintaining local foodways in the face of industrialized agriculture and other extractivist inroads. But indigenous struggles have their special features, though they too map on to the larger agenda of ecosovereignty, as we will see in the next chapter.

References

Altieri, Miguel A., Fernando R. Funes-Monzote, and Paulo Petersen. 2012. "Agroecologically Efficient Agricultural Systems for Smallholder Farmers: Contributions to Food Sovereignty." *Agronomy and Sustainable Development* 32: 1–13.

Altieri, Miguel A., and Victor Manuel Toledo. 2011. "The Agroecological Revolution in Latin America: Rescuing Nature, Ensuring Food Sovereignty and Empowering Peasants." *Journal of Peasant Studies* 38 (3): 587–612.

Bello, Walden, and Mara Baviera. 2010. "Capitalist Agriculture, the Food Price Crisis and Peasant Resistance." In *Food Sovereignty: Reconnecting Food, Nature and Community*, edited by Hannah Wittman, Annette Desmarais, and Nettie Wiebe. Halifax, NS: Fernwood Publishing.

Bernstein, Henry. 2014. "Food Sovereignty Via the 'Peasant Way': A Sceptical View." *Journal of Peasant Studies* 41 (6): 1031–63.

Borras, Saturnino M., and Jennifer C. Franco. 2010. "Toward a Broader View of the Politics of the Global Land Grab." *ICAS Working Paper Series 001*.

Borras, Saturnino M., Philip McMichael, and Ian Scoones. 2011. *Biofuels, Land, and Agrarian Change*. London: Routledge.

Cafaro, Philip, and Eileen Crist, eds. 2012. *Life on the Brink: Environmentalists Confront Overpopulation*. Athens: University of Georgia Press.

Carr, Edward R. 2015. "Political Ecology and Livelihoods." In *Routledge Handbook of Political Ecology*, edited by Tom Perreault, Gavin Bridge, and James McCarthy. London: Routledge.

Chambers, R., and G. Conway. 1992. *Sustainable Rural Livelihoods: Practical Concepts for the 21st Century.* Sussex: Institute of Development Studies.
Chappell, Michael Jahi, and Liliana A. LaValle. 2011. "Food Security and Biodiversity: Can We Have Both? An Agroecological Analysis." *Agriculture and Human Values* 28: 3–26.
Crist, Eileen. 2019. *Abundant Earth: Toward an Ecological Civilization.* Chicago, IL: University of Chicago Press.
Dahbour, Omar. 2013. *Self-Determination Without Nationalism: A Theory of Postnational Sovereignty.* Philadelphia, PA: Temple University Press.
Dasmann, Raymond. 1989. "Towards a Biosphere Consciousness." In *The Ends of the Earth: Perspectives on Modern Environmental History*, edited by Donald Worster. Cambridge: Cambridge University Press.
Davis, John Emmeus, ed. 2010. *The Community Land Trust Reader.* Cambridge, MA: Lincoln Institute of Land Policy.
Davis, Mike. 2005. *Planet of Slums.* London: Verso.
Fairbairn, Madeleine. 2010. "Framing Resistance: International Food Regimes and the Roots of Food Sovereignty." In *Food Sovereignty: Reconnecting Food, Nature and Community*, edited by Hannah Wittman, Annette Desmarais, and Nettie Wiebe. Halifax, NS: Fernwood Publishing.
Gadgil, Madhav, and Ramachandra Guha. 2000. *The Use and Abuse of Nature.* Delhi: Oxford University Press.
Gomiero, Tiziano, David Pimentel, and Maurizio G. Paoletti. 2011. "Environmental Impact of Different Agricultural Management Practices: Conventional Vs. Organic Agriculture." *Critical Reviews in Plant Sciences* 30: 93–124.
Gray, John. 2003. *Straw Dogs: Thoughts on Humans and Other Animals.* London: Granta Books.
Le Billon, Philippe. 2015. "Environmental Conflict." In *Routledge Handbook of Political Ecology*, edited by Tom Perreault, Gavin Bridge, and James McCarthy, 598–608. New York: Routledge.
MacLaren, Chloe, Andrew Mead, Derk van Balen, Lieven Claessens, Ararso Etana, Janjo de Haan, Wiepie Haagsma, O. Jäck, T. Keller, J. Labuschagne, and Å. Myrbeck. 2022. "Long-Term Evidence for Ecological Intensification as a Pathway to Sustainable Agriculture." *Nature Sustainability* 5: 770–79.
Martinez-Alier, Joan. 2002. *The Environmentalism of the Poor: A Study in Ecological Conflicts and Valuation.* Cheltenham: Edward Elgar.
Martinez-Alier, Joan. 2023. *Land, Water, Air, and Freedom: The Making of World Movements for Environmental Justice.* Cheltenham: Edward Elgar.
McDougall, Robert, Paul Kristiansen, and Romina Rader. 2019. "Small-Scale Urban Agriculture Results in High Yields But Requires Judicious Management of Inputs to Achieve Sustainability." *Proceedings of the National Academy of Sciences* 116 (1): 129–34.
McMichael, Philip. 2010. "Food Sovereignty in Movement: Addressing the Triple Crisis." In *Food Sovereignty: Reconnecting Food, Nature and Community*, edited by Hannah Wittman, Annette Desmarais, and Nettie Wiebe. Halifax, NS: Fernwood Publishing.
Monbiot, George. 2022. *Regenesis: Feeding the World Without Devouring the Planet.* London: Allen Lane.
Ondetti, Gabriel. 2008. *Land, Protest, and Politics: The Landless Movement and the Struggle for Agrarian Reform in Brazil.* University Park: Pennsylvania State University Press.
Orr, David W. 1994. "The Effective Shape of Our Future." *Conservation Biology* 8 (3): 622–24.

Patel, Raj. 2009. "What Does Food Sovereignty Look Like?" *Journal of Peasant Studies* 36 (3): 663–706.

Pimentel, David, and Michael Burgess. 2013. "Soil Erosion Threatens Food Production." *Agriculture* 3: 443–63.

Purdy, Jedediah. 2019. *This Land Is Our Land: The Struggle for a New Commonwealth.* Princeton, NJ: Princeton University Press.

Schanbacher, Willam D. 2010. *The Politics of Food: The Global Conflict Between Food Security and Food Sovereignty.* Santa Barbara, CA: Praeger.

Shiva, Vandana. 2020. *Reclaiming the Commons: Biodiversity, Indigenous Knowledge, and the Rights of Mother Earth.* Santa Fe, NM: Synergetic Press.

Smaje, Chris. 2023. *Saying No to a Farm-Free Future: The Case for an Ecological Food System and Against Manufactured Foods.* White River Junction, VT: Chelsea Green.

Smil, Vaclav. 2000. *Feeding the World: A Challenge for the Twenty-First Century.* Cambridge, MA: Massachusetts Institute of Technology Press.

Vince, Gaia. 2022. *Nomad Century: How to Survive the Climate Upheaval.* London: Allen Lane.

Windfuhr, Michael, and Jennie Jonsén. 2005. *Food Sovereignty: Towards Democracy in Localised Food Systems.* Bourton-on-Dunsmore: ITDG Publishing.

Wittman, Hannah. 2010. "Reconnecting Agriculture and the Environment: Food Sovereignty and the Agrarian Basis of Ecological Citizenship." In *Food Sovereignty: Reconnecting Food, Nature and Community*, edited by Hannah Wittman, Annette Desmarais, and Nettie Wiebe. Halifax, NS: Fernwood Publishing.

Wittman, Hannah, Annette Desmarais, and Nettie Wiebe. 2010. "The Origins and Potential of Food Sovereignty." In *Food Sovereignty: Reconnecting Food, Nature and Community*, edited by Hannah Wittman, Annette Desmarais, and Nettie Wiebe. Halifax, NS: Fernwood Publishing.

Wolf, Eric. 1969. *Peasant Wars of the Twentieth Century.* New York: Harper & Row.

Wolf, Eric. 1982. *Europe and the People Without History.* Berkeley: University of California Press.

Wolford, Wendy. 2010. *This Land Is Ours Now: Social Mobilization and the Meanings of Land in Brazil.* Durham: Duke University Press.

Zerbe, Noah. 2013. "The Quinoa Challenge (and Other Food Dilemmas)." *Global Food Politics*. https://globalfoodpolitics.wordpress.com.

2

INDIGENOUS AND PLACE-BASED RIGHTS

Indigenous Peoples' Movements

In the contemporary struggle against extractivism, the most entrenched and therefore effective opposition has come not from environmentalist organizations per se, but from groups expressing the "environmentalism of the poor" (Guha and Martinez-Alier 2000). Principal among such groups and peoples are, on the one hand, as we have seen, farmers, fishers, and hunter/gatherers fighting for sovereignty over the lands and waters necessary for their subsistence. On the other hand, resistance has also been manifested by "indigenous" peoples claiming rights over ancestral homes—lands they have traditionally inhabited.

The history of indigenous, native, and tribal peoples' resistance to extractivism has taken on a characteristic pattern. A first stage is one of treaties signed by colonizing and native peoples, recognizing certain relations from a position of at least diplomatic equality. A second stage upsets these arrangements, rejecting the idea of equitable relations between peoples, and subordinating indigenous peoples as colonial subjects. This stage was often the result of the fact that the "balance of power shifted in favor of immigrant peoples with a growing settler population, increasing military power, and the decimation of indigenous populations through diseases of European origin" (Niezen 2003, 29). The history of North American colonization, among other examples, is a clear case of such a two-stage process of disenfranchisement (Hämäläinen 2022).

Today, this means that still-existent indigenous populations often live on the borders of extractivist economies, where the plundering of natural goods has not yet proceeded very far. The resistance of indigenous peoples to the extension of resource extractions has "arisen out of the shared experiences of marginalized

DOI: 10.4324/9781003568155-4

groups facing the negative impacts of resource extraction and economic modernization" (Niezen 2003, 9). They directly face threats to their own livelihoods because they "live on the *geographic frontier* of the advancement of the Western development juggernaut" (Mander 2006, 226; italics added).

This fact—that "indigeneity" is characteristic of a *geographically* specific phenomenon—will be used to characterize, define, and to some extent, universalize the indigenous rights movement. While, as we will see, indigenous *rights* have been largely the concern of international jurists, indigenous *identities* per se have been discussed primarily by anthropologists interested in developing a definition of indigeneity. Yet, such definitional attempts have been sharply criticized for their indeterminacy and lack of specificity, in many cases by anthropologists themselves. At the same time, the efforts of jurists to articulate an understanding of (indigenous) rights that can cover the wide array of cases pertinent to international legal documents have yet to succeed. It will be maintained here that both of these attempts have failed due to the lack of understanding that indigenous rights is an inherently normative concept, one that articulates certain values—values that are, at the same time, tied to a geographical, more than an anthropological, notion of indigeneity.

The fundamental value that underlies indigeneity is that of the importance of places, including lands and territories. This means that indigenous rights are an important, but not exclusive, instance of a more general concept, that of "place-based rights." Such rights, founded on *attachments to place*, and given a functional definition as being concerned with a specific relation to land, territory, and resources, will provide a strong underpinning for both immediate struggles and long-term claims to sovereignty and self-determination. It is therefore the purpose of this chapter to show how indigenous rights plays a crucial role in *expanding* what has sometimes been regarded as a narrowly defined conception of property rights.

So we need to consider both the nature of indigenous rights (as potentially legal entitlements) and the definition of indigenous peoples (which is required in order to be clear about legitimate claimants of rights). But before we do this, consideration should be given to how the notion of indigenous rights differs from the standard emphasis on states as the principal claimants of collective rights of this kind. The most important context for thinking about the relation of indigenous peoples to states is the twentieth-century struggle for colonial liberation and the self-determination of peoples. A common understanding of this struggle is that liberation movements sought the establishment of independent states in the colonies of the European empires. Yet, indigenous movements challenge this definition of self-determination. For them, states are not the best means of self-determination and liberation: decolonization means liberation from the state *system* as a whole: "The fulfillment of the self-determination goals of indigenous peoples would constitute a completion of the process of decolonization through mechanisms that would *not* involve the creation of new nation-states" (Niezen 2003, 194; italics added).

There are a couple of reasons for this lack of interest in statehood and political independence on the part of indigenous movements. For one thing, states tend to favor majority populations, rather than minorities, especially minorities that have radically different customs, identities, and practices. While this can apply to, for example, racial or ethnic minorities in some instances, it applies particularly to native, tribal, and indigenous peoples who may *refuse* to participate in the construction of modern states and markets, or may regard states—of whatever origin—as institutions bent on imposing alien ways of life. From an indigenous point of view, self-determination does not involve statehood because states have a (self-)interested view of the allocation of rights, benefits, and resources (Niezen 2003, 196). In addition, modern territorial states invariably have uses for the lands and territories of "undeveloped" regions and places. This almost always pits them against indigenous populations that seek to continue to live in traditional ways: "The development goals of newly independent states have proven to be at odds with the subsistence viability and cultural survival of those living in frontier environments" (Niezen 2003, 196).

Indigenous peoples do not, therefore, generally seek to found states, as they are understood in the tradition of European philosophy and law (Alfred 1999, 56). When they do come to advocate self-determination, and ultimately sovereignty, it is to "seek only a limited form of autonomy, not independence; the goal relates only to powers of self-government within a framework of constitutional law and authorities delegated by the state" (Alfred 1999, 57).

In the end, indigenous peoples usually want an "honorable relation with states," in the sense that such peoples will not challenge state authority, *if* states respect indigenous rights to lands, territories, and cultural practices (Niezen 2003, 18). But this means that indigenous peoples increasingly must, in order to protect their ways of life, and places in which to practice them, seek recognition of their rights by local states, as well as global institutions. It is to this task that a number of jurists have developed considerable effort over the last two generations.

Dilemmas of Indigenous Rights Claims

The history of indigenous rights claims in international law stretches from the first considerations of indigeneity in the 1960s to the present. But the institutional history can best be regarded as spanning two documents, one from 1989, and a second from 2007. The first was a "convention" of the International Labor Organization (ILO) concerning "Indigenous and Tribal Peoples in Independent Countries"; the second is the UN Declaration on the Rights of Indigenous

Peoples. The ILO. Convention was ratified by only a few states; the UN Declaration by many more. But neither had treaty status, though they were widely disseminated and referenced in most discussions of indigenous rights. In addition, the World Bank, influential because of its financial importance for "developing" countries, published several reports on indigenous rights, which were notable for their clear and thoughtful consideration of problems of definition. Perhaps most important, but to be considered later, is the recognition that such rights have been accorded in a number of state constitutions.

Three issues have been paramount in the extensive discussions among jurists concerning indigenous rights: the rights to be accorded to claimants, the nature of the claimants themselves (especially whether or not they constituted "peoples"), and whether there was anything distinctive about indigenous rights, as compared with other collective rights claims. These issues are distinct from, though bearing upon, the central problem of defining indigeneity, a problem more commonly addressed by anthropologists than jurists, and one to be considered next. The legal issues in particular stem from the fact that there is no obvious referent with regard to indigenous people(s). As Bengt Karlsson puts it, "'Indigeneity' is a traveling discourse that has emerged and developed in dialogue with various social movements and non-indigenous actors, and not something that reach[es] us straight from the mud-hut, bush, or wherever one is to locate the 'authentic' tribal spokesperson" (Karlsson 2006, 55). This means that, for juridical discussions, there is no necessary or stable framework that can limit or constrain the various definitions of rights claims. For Benedict Kingsbury, one of the most thorough commentators on the history of the discussion, clarity can only be achieved through "constructivist" means, that is, by "embodying a continuous process in which claims and practices in numerous specific cases are abstracted in the wider institutions of international society, then made specific again at the moment of application" (Kingsbury 1998, 415).

The least controversial part of the legal treatment of indigenous peoples is the actual rights that can be claimed. Already in the ILO Convention, it was clear that they involved essentially three things: [1] protection, and perhaps enhancement, of the cultural integrity of indigenous groups, [2] access to, and preferably, ownership of, ancestral lands and resources, and [3] political self-determination of the indigenous, in distinction from existing states (Anaya 1991, 15ff). In each case, there are some controversies as to what precisely is involved. In the first case, whether indigenous groups require positive assistance from states—for instance, in the establishment of separate schools—is unclear. Perhaps it is sufficient to allow the use of native languages, the practice of indigenous religions, and the toleration of private school systems in order to satisfy this right. In the second case, the extent to which indigenous groups actually own and control certain lands is in question. Furthermore, the definition of what constitutes "ancestral" or traditional territories is to be determined. Finally, the extent of self-determination, up to and including sovereignty rights such as separate legal systems or even political independence (as a result of secession from existing states), is a matter of considerable debate. Except in the last instance, these matters do not seem to provide grounds for the actual rejection of indigenous rights; rather they involve definitional matters that might require additional financial commitments by states or modifications of existing property rights.

But much of the controversy around indigenous rights concerns the definition of who has valid claims to these rights. This in turn is determined by two things: how the indigenous in general are to be characterized, and which specific characteristics denote a group as indigenous. Are the indigenous best characterized as groups, communities, people, or people*s*? While all of these terms have their advantages and disadvantages, the most persistent controversy has concerned which of the last two terms is most applicable.

The reason for this is the history of the use of the term "peoples" in international law. While the term has famously never been defined in legal instruments, that has not prevented its frequent usage in numerous declarations and other documents over the past fifty years or so. Of particular importance in this regard are a series of UN declarations that defined the right to self-determination of colonized peoples, especially UN General Assembly Resolution 1514 (XV), "Declaration on the granting of independence to colonial countries and peoples" (1960). This was followed by numerous subsequent resolutions in the early 1960s (and later) on specific instances of self-determination for colonial and semi-colonial peoples (e.g., Algeria, Palestine) (Dahbour 2003, 63–65). In these documents, "peoples" were used to refer to the subject populations of the European colonies and former colonies. To now refer to indigenous groups using the same terminology might imply that they had the same rights as the peoples recognized previously in these international legal instruments.

"People," in contrast, refers to conglomerations of individual persons, or "populations" of discrete individuals. For instance, in the 1992 Declaration on Environment and Development, the result of a meeting in Rio de Janeiro, there was recognition of "indigenous people and their communities," but not indigenous peoples as such (Barsh 1994, 46). The following year, at the UN World Conference on Human Rights, this issue came to the fore with the "battle of the 's'"—that is, a fight over whether indigenous people should be accorded the name of peoples (Barsh 1994, 51–52). It proved impossible at that time to achieve consensus on this.

The reason was that using peoples to characterize the indigenous seemed to equate them with the peoples inhabiting independent states—and therefore, to implicitly recognize indigenous peoples as having the same rights of self-determination, sovereignty, and so on as such independent peoples already had. Needless to say, this seemed to suggest that indigenous peoples, were they to be so recognized, could claim rights of self-governance, up to and including secession from existing states, at least under some circumstances. This was clearly at that time too much to accept for a number of states, in particular several Asian states such as India and China. By the time of the issuance of the UN Declaration in 2007, the terminology of indigenous peoples had been largely accepted; but of course, the declaration has no real force as law, despite its subsequent ratification by over a hundred states.

This now brings us to the major issue in the legal discussion of indigeneity—the specific characteristics that constitute indigenous peoples (the anthropological discussion, which is somewhat distinct, will be considered next). Two questions in particular constitute points of controversy: [1] do indigenous peoples constitute the original inhabitants of a territory, or only the "native" inhabitants? And [2] are indigenous peoples always the subjects of a colonial power at some point, or is this not a necessary feature of indigeneity? A quick survey of some important statements about indigeneity will show some ways in which these controversial points have been treated.

An influential early attempt at definition was made by José Martinez Cobo, in 1987, in a report to the UN. He defined indigenous communities as those that: [1] had an historical continuity with precolonial societies, [2] existed currently as distinct communities, [3] were nondominant parts of a larger society, and [4] showed a determination to preserve their way of life and distinctness as a community (Niezen 2003, 20). This attempt at an expansive and comprehensive definition can be contrasted with attempts in the 1990s to find a more restrictive, but also more precise, definition. In 1999, a UN special rapporteur, Miquel Alfonso Martinez, issued another report sharply at variance with Martinez Cobo's. In it, Alfonso Martinez claimed that indigeneity is a condition subject to "organized colonization, by European powers, of peoples inhabiting, since time immemorial, territories on other continents" (Karlsson 2006, 60). This would effectively rule out many Asian peoples who might be considered, by more expansive criteria, to be indigenous. Finally, a World Bank report from 1991 gives a more functional definition, which sidesteps the issue of colonization: "The terms 'indigenous peoples,' 'indigenous ethnic minorities,' 'tribal groups,' and 'scheduled tribes' describe social groups with a social and cultural identity distinct from the dominant society that makes them vulnerable to being disadvantaged in the development process" (Kingsbury 1996, 23).

In the end, despite reservations about it, the term "peoples" was generally accepted as the entity to which "indigenous" referred. While concepts of "groups" or "communities" were, in some instances, offered as substitutes for peoples, these were, especially by indigenous representatives themselves, judged insufficient. This was largely because, as Kingsbury noted, they lacked any connection to the idea of self-determination that was, whatever its meaning, increasingly thought to be an important aspect of indigeneity:

> The international concept of "indigenous peoples" connotes emphasis on self-determination and the role of groups in decisions affecting them, respect for different cultures shaped over long periods of history, recognition of special relations with land and territory and unique knowledge about their use and management, and awareness of the disastrous consequences for these peoples of many prior policies of states and international institutions.
>
> *(Kingsbury 1998, 453)*

Nevertheless, this acceptance of "indigenous peoples" did not end the controversies over its meaning and reference. In particular, the restrictive definition favored by the Indian and Chinese governments, powerful actors in forums concerned with defining indigeneity, continued to be favored by many newly independent states (Kingsbury 1998, 433). For them, the condition of "saltwater settler colonialism" was a necessary aspect of indigeneity (Kingsbury 1998, 434). That is to say that indigenous peoples were essentially and only "first peoples" who had been colonized, displaced, or decimated by European settler states or populations. This was the case in much of North and South America, and certain parts of Africa and Oceania, but *not* in the main civilizational regions of Asia.

Thus, where confrontations between indigenous peoples and European settler populations created conditions of disenfranchisement and subordination, indigeneity was an apt concept; otherwise, it was not. Certainly, as Kingsbury suggests, the concept was less *easily* applicable to situations in which this confrontation was absent (Kingsbury 1998, 447). But it was also the case that "it seems unlikely that the PRC's objection to application of the concept of 'indigenous peoples' to China is an objection *in principle*" (Kingsbury 1998, 449). More likely is the fact that, were the expansive definition accepted, Chinese expansion into, and subordination of, Manchuria, Mongolia, Tibet, and Xinjiang might be viewed as violations of indigenous self-determination. Furthermore, in the Indian case, were, for instance, relations between Dravidian and Indo-Aryan groups to be viewed as a case of confrontation between indigenous and non-indigenous peoples, the same perspective might be applied.

While the restrictive definition has the advantage of precision and clear applicability, the expansive definition seems to be applicable to instances in which many of the characteristics of indigeneity generally occur. To summarize, these characteristics are the following: [1] historical continuity with the area, perhaps predating its colonization, [2] contemporary subordination to, or domination by, a majority population, [3] connection to the area, identified as a homeland or ancestral territory, and [4] self-identification as a distinct group, people, or "nation," perhaps on the basis of certain cultural or linguistic characteristics (Kingsbury 1998, 453–54). Further consideration of these features will be given when we discuss the anthropological literature. However, it should be clear that both definitions have something to be said for them, and the legal discussion does not have the means for resolving the contradiction between them; for this, we will have to turn elsewhere. It is also important not to forget the "functional" definition offered by the World Bank, where indigenous peoples' distinctiveness "makes them vulnerable to being disadvantaged in the development process." This is an awkward way of indicating that indigenous groups may not want to participate in such a process, and it may be *that* which makes them distinctive. But we will return to this point later.

Finally, it is worth considering the liberal critique of indigeneity, which seeks to accentuate the definitional problems with the concept in order to show that it plays no essential role in the articulation of injustices or inequities between groups. This criticism can be found in various places, but a key statement is given by Jeremy Waldron in his 2007 article, "Why Is Indigeneity Important?" (Waldron 2007; cf. Buchanan 2004, 408–15). Waldron regards arguments for indigenous rights as being of two kinds, one arguing for a right of "first occupancy," the other for a right of "prior occupancy." In neither case does invoking indigeneity add anything essential to earlier philosophical arguments for such rights—arguments given by, among others, John Locke and Hugo Grotius, in the early modern era. In fact, contemporary indigenous rights arguments, according to Waldron, subtract nuance and conditionality from these rights, making them blunt instruments unable to account for the situations in which various considerations would indicate that the rights do not apply. In the case of first occupancy rights to property, for instance, Locke made them conditional on "labor and cultivation," not an absolute right of appropriation as such (Waldron 2007, 30). Furthermore, Locke stipulated a "proviso," where the right only applies if there is "enough and as good left for others"—in other words, a distributive condition taking into account the needs of all affected by their exclusion from certain lands or properties (Waldron 2007, 36). Waldron's point is that a right of first occupancy can be rendered illegitimate, if it is unconditional. Are indigenous rights unconditional in just this way? Do neighbors, the needy, and even settlers have claims that must be accommodated? Justice would seem to require an account of why and when they are or are not excluded. A simple "right of first occupancy" does not do this.

The right of "prior occupancy" has a different problem: it is the temporality of the right that is paramount in determining to whom it obtains. Waldron regards this version of indigenous rights as inherently conservative—in fact, a "principle of established order" (Waldron 2007, 33). It seeks to protect the "stability, security, certainty, and peace" of a certain property arrangement (Waldron 2007, 31). But which one? In some cases (Waldron mentions New Zealand), the dominance of a non-indigenous population is itself of some considerable length (since the nineteenth century at least). If prior occupancy is to be protected, this could apply to new patterns (of "established order") if they have lasted awhile. But in other cases (Palestine comes to mind), "If the date of the injustice [e.g., invasion or colonization] and the date of the claimed reversion are sufficiently close," then reversion to prior occupancy may be an "appropriate remedy" (Waldron 2007, 33).

The greater worry is about who is to adjudicate such controversies or conflicts, especially when such rights are indeterminate in meaning, either between first and prior occupancy rights, or between different meanings of indigeneity itself. As Waldron concludes, "There are places in the world where making [prior occupancy] the crucial question is a deadly and vicious ingredient in social and political pathology" (Waldron 2007, 38). If indigeneity contributes to such outcomes—civil war, ethnic cleansing, and so on—then perhaps it is better left out of considerations about property rights, distributive justice, and similar matters.

The discourse of international law is ill-suited to resolving quandaries about the meaning of indigeneity and indigenous rights. The apparent indeterminacy of the key concepts of indigenous people(s) and property rights results in irresolvable dilemmas concerning a definition of the terms that would allow consensus building on the global scale. This, in turn, opens the concept of indigenous rights to philosophical doubts about its utility for specifying what justice between dominant and subordinate groups in multicultural societies could mean. Anthropologists, in attempting to define indigeneity, encounter similar difficulties, as we will see in the next section.

However, there are solutions to the indeterminacy of indigenous rights, though they are to be found neither in law, nor in anthropology, but in concentrating on the geographical settings, and people-land relations, of specifically indigenous peoples. From the standpoint of liberal political philosophy, there is little to distinguish between different collectivist attempts at territorial claims, whether indigenous, ethnonationalist, or otherwise. But, as we will see in Chapters 3 and 4, these types of claims are actually quite distinct. What bothers Waldron and other liberal theorists is that they are not founded on individualist premises, and therefore might provide grounds for violating individual rights to property ownership, alienability of assets, freedom of movement, and so on. But these are not the fundamental concerns for indigenous peoples, however defined. It is preserving attachments to lands, ways of life, and natural goods important for peoples' livelihoods that are paramount. Once it is understood that indigeneity involves a relation between peoples and lands that is unlike other individualist *or* collectivist relations, which seek to distinguish individuals or groups from *other* individuals or groups, this difference becomes clear. It is the geographical relation that *fixes* the terms and identities, and thereby also admits of conditions and limits. These conditions and limits are not a function either of individual rights or national identities, but of human-place relations that are not easily eliminable or alienable (though, of course, extractivism is one way in which they can be degraded or destroyed). Before further defining these relations, however, consideration needs to be given to the extensive efforts, in particular by anthropologists, to define indigeneity apart from the vicissitudes of international law.

Problems With the Definition of Indigenous Peoples

The debate among anthropologists about indigeneity follows a similar but distinct trajectory to that of the jurists. There is no concern, as there is in legal circles, with agreeing upon a definition that can gain the assent of states. But there is more concern about the distinctiveness of indigeneity in comparison with other terms employed by anthropologists, such as tribes, ethnicities, classes, castes, and so on. Furthermore, there is the worry that a term originally designed for some contexts (e.g., North and South America and Australasia) is being misapplied to other contexts (e.g., India and Africa) where it is less applicable. In addition, there is a desire to "let the indigenous speak"—that is, to allow for the possibility of self-identification. All these considerations lead to attempts to specify the identity that indigenous peoples have, rather than the rights they might claim in a global legal or political system.

Some scholars view indigenous peoples on a continuum with other categories such as tribal peoples, ethnic minorities, and ethnonational groups (Niezen 2003, 5). The upshot of this approach is for indigeneity to lose its specificity in relation to these other terms. Aside from facing the question of whether to regard indigenous peoples as original (or first) inhabitants or as inhabitants (immediately) prior to colonization or conquest, there are two criteria that are common in anthropological discussions. The role of the indigenous as a subordinate or nondominant group, in relation either to a colonizing power, a majority culture, or contemporary extractivist regimes is essential to most definitions (Karlsson and Subba 2006, 6). However, this does not necessarily differentiate the indigenous in some contexts from subordinate classes (e.g. peasants) or castes (e.g. Dalits or "untouchables" in South Asia). Also, there seems to be the necessity of some self-identification as having a distinct culture or ancestry. In this case, it is often claims of parentage or locality, rather than distinct practices or beliefs, that determines identity among the indigenous themselves (Niezen 2003, 13).

Among some anthropologists, these attempts at definition have generated considerable skepticism about the use of the term, indigeneity. As André Béteille has written, "The phrase 'indigenous people' is a little like the term 'native' in colonial usage, with the moral significance reversed to some extent . . . . Has the crude anthropological association of race and culture acquired a more refined form in the concept of the indigenous people?" (Béteille 1998, 190). Such critiques have gained force from groups in some regions (e.g., India) who have rejected the terminology of indigeneity in order to self-identify as tribes or castes. This has particular force in South Asia where the term "Adivasi" remains in wide usage:

> [Adivasi] connotes political awareness and the assertion of rights. It is a Sanskrit derivation that interestingly means "original inhabitants." The term was coined in the 1930s as part of the struggle against the intrusion by the colonial government and outside settlers and moneylenders . . . . Being Adivasi . . . "is about shared experiences of the loss of the forests, the alienation of land, repeated displacements since independence in the name of 'development projects.'"
>
> *(Karlsson and Subba 2006, 3)*

The doubts that anthropologists have raised about the concept of indigeneity are complemented by a critique of the concept made by other scholars, including historians, especially in South Asia. While the critique of indigeneity put forward by the Chinese government in various international bodies has been itself criticized as self-serving, as we indicated earlier, the South Asian critique is somewhat different. While still state-based or allied to positions put forward by the Indian government, the worry is somewhat different in this case. Here, scholars have argued that, in many instances, not limited to South Asia, it is impossible to determine either the first inhabitants of regions, or the most ancient ones, since more than one people have inhabited such areas for many centuries. In multicultural societies, it is often impossible to determine the relative antiquity of different groups, and in any case, may be counterproductive, increasing local antagonisms needlessly (Karlsson 2006, 56).

Furthermore, some claim that the idea of indigeneity is a calculated misapplication of the concept, used by some to destabilize or delegitimize postcolonial states (Karlsson and Subba 2006, 7). Advocacy of "cultural rights" for the indigenous is also regarded as a dangerous policy that can undercut the "political rights of citizens within a state" (Baviskar 2006, 38). In addition, the idea that indigeneity connotes a different attitude toward the natural world and various ecosystems is sometimes regarded as threatening the development policies of newly independent states. For all these reasons, indigeneity is regarded by many scholars as an ill-defined and politically problematic concept, especially when applied to areas without recent experience of European colonization.

This, as we will discuss further in the next section, raises the issue of the *value* of indigeneity as a concept denoting particular peoples and their relation to their environments. Indigeneity has often been associated with a distinct view of nature, one not subsumable under a modernizing/industrializing process, such as has been pursued by many "underdeveloped" states. As Victoria Tauli-Corpuz writes, "Many indigenous peoples are still awed by the magic of the cosmos and the mystery of life [while] industrialized culture regards our values as unscientific obstacles to modernization and thus worthy of ridicule, suppression, and denigration" (Tauli-Corpuz 2006, 14–15).

Even under conditions of political struggle for indigenous *rights*, where indigenous (including tribal and Adivasi) peoples must themselves adapt to modern culture and media in order to succeed, such an emphasis on the cultural distinctiveness of indigeneity can be regarded as efficacious. As Amita Baviskar writes,

> The cultural rights argument remains a powerful resource for mobilization and one that is frequently used by adivasis themselves who claim that their culture is organically linked to a particular ecology. Asserting an indigenous identity enables adivasis to claim rights to natural resources.
>
> *(Baviskar 2006, 38)*

While criticisms of the concept of indigeneity show its problematic status as an overgeneralization from certain regions to the world as a whole, the concept retains its power as a means of arguing against development policies and state-building. When such policies have deleterious environmental and social effects, as they often do, especially for tribal and other rural peoples, indigeneity can provide a powerful argument against the extraction of resources from still-intact ecosystems. But it remains a question of how best to define and operationalize the concept so as to effectively counter extractivism.

Indigenous Rights as a Normative Concept

The attempt by jurists to "construct" a concept of indigeneity that satisfies all major participants in international forums concerned with the issue is bound to fail (Kingsbury 1998, 414–15). There are irreconcilable interests that will largely thwart such attempts. At the same time, the efforts of anthropologists to formulate a meaning for indigeneity that would cover a wide variety of cases, including those not subject to "saltwater colonialism," founders on the understandable reservations of scholars who see alternative concepts—native, tribal, Adivasi—as more specifically applicable to particular regions and cultures with which they are concerned.

Breaking this impasse—whether or not to use indigeneity as a widely applicable concept, and how to understand the rights that follow from it—requires a disciplinary shift. The first aspect of this is to understand that indigeneity, like a number of other social scientific concepts, has an irreducible normative component. There are reasons to use the concept—as well as reasons not to do so—and these must be highlighted and factored into whatever definition is ultimately employed. Yet, some scholars who recognize this nevertheless insist that the normativity of indigenous rights does not necessarily settle questions about its meaning, since norms are variable and subjectivistic. Here the relativist biases of anthropology come into play; philosophers thinking about the question of normativity are more likely to argue for the possible objectivity of values that may be applied (and this is the position adopted here). For instance, Benjamin Gregg, who rightly emphasizes the normativity of the concept, nevertheless does not see why this has the potential to resolve some of the quandaries about meaning and definition:

> The term *indigeneity*, invested in any given usage with particular value commitments, is not value-neutral . . . it is itself historically contingent, culturally perspectival, and morally relative. Hence the political question: whose values should determine the content of any particular definition of an indigenous people?
>
> *(Gregg 2020, 102)*

Framing the problem in this way is an undoubted advance over previous discussions. However, without an adequate conception of rights, any specific definition can seem functionally equivalent to any other. But if rights are understood in their proper signification, this may provide resources for narrowing and specifying the range of adequate meanings of indigenous rights. The first task is to be clear about the limitations of rights discourse as applied to indigeneity. Despite two generations of activity in defining and espousing indigenous rights, there has been little success in specific localities in terms of achieving protection for indigenous peoples and their livelihoods (Corntassel 2008, 113). Some have argued that the reason is that legalistic strategies of (human) rights-oriented legislation on the global scale inevitably orient indigenous peoples toward accommodation with existing states that presumably would recognize and enforce such rights. Such states often have their own reasons for limiting the scope of such rights, and selectively enforcing them in relation to other goals and priorities, such as economic development, resource extraction, and population resettlement. For this reason, "Strategies that invoke existing human rights norms and that solely seek political and legal recognition of indigenous self-determination will not lead to a self-determination process that is sustainable for the survival of future generations of indigenous peoples" (Corntassel 2008, 108).

A different understanding of rights is required before they can be serviceable for the long-term survival and resilience of indigenous peoples. The beginning of such an understanding is the recognition of the correlativity of rights and responsibilities. This idea, rooted in philosophical critiques of individually oriented rights discourse by Hegel, Hart, and others, claims that only rights concepts that correlate with concepts of responsibilities as well are able to achieve their goals (Hegel 1991; Hart 1983). Rights, as Hegel was the first to emphasize, always imply duties. In the indigenous case, it is, therefore, the responsibilities of indigenous peoples that must be discharged for the rights to have meaning (Corntassel 2008, 116). What duties or responsibilities? Above all, the rights must correlate with duties derived from "indigenous relationships to the natural world" (Corntassel 2008, 108). What makes indigeneity unique is not, in other words, a particular relationship to, or difference from, other cultures, groups, or nations, but a relationship with the local natural and ecosystem contexts within which indigenous peoples live. Only if this relationship is protected, maintained, and strengthened can indigenous peoples achieve a measure of self-determination that is generally the goal of assertions of indigenous rights in the first place.

Rights held *against* other peoples must therefore correlate with responsibilities held *for* ecosystems which the indigenous require in order to *be* indigenous (to a locality, area, or region). A term for this might be "sustainability"; self-determination, in other words, must correlate with sustainability. This is the core *value* that defines indigeneity; and it is not "culturally relative." It is present, as we will see in more detail further, in the very meaning of indigeneity, since absent some such relation to the natural world, indigeneity loses specificity and meaning, as opposed to the various other forms of culturally local and/or subordinate identity (e.g., ethnicity, nationality, minority status).

It may be the case, as Kingsbury claims, that, as a descriptive matter, indigenous peoples may vary widely in their "practices and attitudes relating to resource exploitation and environmental maintenance" (Kingsbury 1998, 440). The question is whether nonextractive practices and beliefs are in any way inherent in, or definitive of, indigenous ways of life. Kingsbury does admit that debates about indigeneity have been affected by "conflicts over land, forests, mineral resources, fishing rights and other valuable natural resources" (Kingsbury 1998, 444). If the relation of indigenous peoples to natural environments is completely elective, why would this be the case? In fact, while indigenous groups can be, and have been, "coopted" by states and corporations seeking resource extraction, this should not be surprising given the stakes. This means, however, that, in not discharging their *responsibilities* to the ecosystems they inhabit, indigenous peoples may be putting their own claims to "special rights" at risk.

Evidence for the special responsibility of the indigenous toward nature comes from many sources. One already noted is the 1991 World Bank report that acknowledged indigenous peoples as distinct on account of their special relation to the "development process." Such a view caused the World Bank to withdraw support from the Sardar Sarovar dam project on the Narmada River in Malwa (India), which was flooding tribal areas (Kingsbury 1998, 444). This development project was also the occasion for one of the largest anti-development protests in post-independence India.

In the same period, the 1992 Rio Declaration (of the Summit Conference on Environment and Development) recognized that there is a "unique status for indigenous peoples that is justified by their traditional reliance on and sustainable management of renewable resources" (Barsh 1994, 48). Studies of indigenous peoples in the Amazon River Basin have shown that, "What all these people have in common is a good environmental record based on low-impact techniques and a stake in retaining or regaining control of the territory they exploit (e.g., for rubber or nuts) (Carneiro da Cunha and de Almeida 2000, 317). More recently, negotiations toward a UN biodiversity treaty have been criticized for not including indigenous peoples, since the latter, as a result of their inclusion *within* ecosystems, have been shown to practice among the most effective forms of preservation of forests and fisheries (Sengupta, Einhorn, and Andreoni 2021).

It is reasonable, therefore, to regard *ecological sustainability* as the value underlying and defining indigeneity. Such a value, while having different meanings in different regions and locales, seems to be essential to the connections between indigenous groups and ecosystems, and is not "relative" to certain cultural identities, or to the policies that indigenous groups may (or may not) adopt. This point can serve to highlight the descriptive and normative aspects of the concept. Certainly, it is necessary to begin defining indigeneity by focusing on the ancestral ties that are claimed, and descriptively validated, by indigenous groups. That descriptive validation means that such ties cannot merely be claimed, as in contentions by some ethnonational groups that they have connections to particular places (e.g., modern Greeks to the Greek peninsula, or Jews to Palestine). There must be some *living* connection between indigenous groups and their lands—either in terms of actual occupancy or recent memory, or some combination of them. Given the controversies about expansive or restrictive definitions of indigeneity, it is important to consider how liberally to take such connections, and the claims that may follow from them.

The fact that ancestral claims constitute a necessary, but not sufficient characterization of indigenous identity makes an expansive definition acceptable. While the relatively few cases of indigenous claims in the face of settler colonialist dispossession obviously qualify, it is other, more ambiguous cases that give pause. But if the initial assertion of ancestral ties is regarded as only the beginning of a claim to indigenous rights, not the whole justification, the dangers of an expansive definition will lessen. It is the normative aspect of indigeneity that will do the work in sharpening and narrowing the specifications of indigeneity. In determining *which* ancestral ties count *where* and for whom in making indigenous rights claims that the normative criterion suggested earlier—that of ecological sustainability as an essential desideratum of indigenous rights—comes to bear. For it is only those groups prepared to forego exploitative, extractive, capitalizing, and fungible relations to ancestral lands who will ultimately be justified in claiming rights to occupancy of, and control over, those lands. The reason for this lies in the particularly close relation between land and ancestry in the determination of indigenous identities.

Indigeneity as the Interplay of Land and Ancestry

Regarding indigeneity as a matter of specifying a cultural identity of a certain type is an exercise in frustration, as the debates among ethnologists about the differences between indigenous, tribal, ethnic, national, and other groups have shown. Unless some external referent to identity formation can be found, self-identification, racial stereotyping, and other unhelpful markers of identity will dominate. By shifting from an anthropological/ethnological to a geographical/ecological framework, such an external reference point can be found. That point is the importance of land and territory in the formation of indigenous cultures. Only if indigeneity is understood to be a particular relation to land, held by those peoples with a verifiable claim to ancestral ties to that land, can the impasses of legal and anthropological definitions of indigeneity be avoided. The rights claims that could follow from such an understanding of indigeneity would be validated, not merely by the existence of ancestral connections but also by the demonstrable willingness of indigenous peoples to assume their responsibility for the sustainable management of their lands for present and future generations. These would be the necessary and sufficient conditions for such a rights claim.

This solution is there for all to see. One of the constants of debates about indigeneity has been a reference to the importance of territory for indigenous peoples. Sometimes this importance is denied, even as the evidence for it is presented. For instance, an early report "recognized that indigenous peoples tend to be characterized by their maintaining special relationships with their 'ancestral territories.'" The ILO Convention from 1989, despite not including "any geographical factor in its definition of 'indigenous,'" affirmed this "continuing relationship between indigenous peoples and their ancestral territories" (Daes 1996, 16). But obviously, this was an essential "geographical factor!"

There are two elements of this relationship that bear noting: the relation of land to ecological sustainability and the relation of territory to cultural identity. The identification of sustainability as the core value of indigeneity implies an important connection between the maintenance of land as a conjunction of healthy soils, flourishing flora and fauna, and sustainable human habitation, and the presence of indigenous peoples. From the perspective of indigeneity, this may seem to imply the right to own lands as collective properties. But in fact it is the more important part of this connection that indigenous peoples have duties to the land (and waters)—to maintain it in a healthy and resilient condition.

Consider the following statement from Victoria Tauli-Corpuz, chair of the UN Forum on Indigenous Issues:

> There is no question that we have inherent and prior rights to our territories and lands, since our ancestors occupied these places long before nation-states

> came into existence . . . . *We also have been the protectors of these places for centuries and longer, keeping the resources and ecosystems intact.*
>
> *(Tauli-Corpuz 2006, 216; italics added)*

While she regards indigenous rights as essentially the rights of first occupants to lands and territories that they have occupied for centuries, it is the second aspect of their role as land and water "protectors," that should be regarded as equally important. Waldron notes that rights of first occupancy must be conditional since it cannot be *assumed* that there have been no other occupants since who could not make similar claims. It is obvious that the rights and needs of subsequent inhabitants have no claim at all to such lands, any indigenous claims must take account, in some way, of the relative validity of any subsequent interests in or needs for the lands that more recent inhabitants may have. Furthermore, there may be cases in which indigenous peoples fail to assume their responsibilities to maintain their lands sustainably for their own people (and others who may have some claim to it as well), in favor of extractivist practices or simply neglect. In that event, they would lose their right to *exclusive* occupancy of and control over the land, *even if they could claim ancestral connections to that land.*

In order for this conception of indigenous rights to work, there would need to be some essential connection between lands and territories and the cultural identities of indigenous peoples. This connection has been noticed by both jurists and scholars of indigeneity, but not given the critical role that it deserves. Kingsbury notes the "central importance of land and territory to group identity and culture," but goes on to list many other factors as being of supposedly equal significance (Kingsbury 1998, 437). This connection was already enshrined in international documents by 1994 since the International Conference on Population and Development (the Cairo conference) had affirmed the "centrality of land tenure systems and ecological knowledge to cultures of indigenous peoples" (Daes 1996, 16).

From the perspective of indigenous peoples themselves, this connection is quite clear. It is what might be thought of as the connection between the "spiritual" and the "material" elements of indigenous practices. As Taiaiake Alfred writes, such practices are based upon a "spiritual connection with the land established by the Creator [which] gives human beings special responsibilities within the areas they occupy as indigenous peoples, linking them in a 'natural' way to their territories" (Alfred 1999, 61). Tauli-Corpuz succinctly encapsulates the view by stating, "For indigenous peoples, keeping our territory or ancestral land is the most important thing. *This is what determines our identity*" (Tauli-Corpuz 2006, 15; italics added).

The anthropological attempt to fix indigenous identities by focusing on various characteristics rooted in kinship, or other features of human cultures has, as we have seen, failed in its task. Rather, we should shift our thinking to the geographical determinants of indigeneity associated with attachments to lands and territories. Peoples and lands, as I will argue more generally in Chapter 4, are codetermining features of indigeneity, or indeed, of any close association between peoples and the physical contexts in which they find their places and homes. In the specific context of indigeneity, Brian Burkhart has made this point in the following way:

> People do not create territory . . . territory creates a people. If people create territory, the territoriality exists as a human convention . . . that floats free from the land itself. This is not how territory works through epistemic practices of locality or in the narratives of indigenous people regarding their relationship to land. I do not choose the land, the land chooses me. The land is what makes me who I am. It is not a convention.
>
> *(Burkhart 2019, 90)*

In fact, Burkhart argues that the very contemporary obsession with (cultural) identities (including indigeneity) needs to be historically located in the development of modern colonial empires, which invented ways of categorizing their (new) human subjects. Subsequent adoptions of "identity politics" by anticolonial actors and thinkers only mimicked such inventions, much as the creation of postcolonial states resulted in often failed imitations of parts of the European state system:

> "the concept of identity is itself historically produced" (Bracken 2014), and produced under the delocalizing force of modernity/coloniality that frames the production of . . . particular identities . . . . The creation of identity as a way of framing subjectivity is part of the attempt to manifest being through delocality . . . and operates in service of erasing Indigeneity, not as just another delocalized identity but as a seat of locality that must be obscured in the process of colonializing the land.
>
> *(Burkhart 2019, 81)*

Indigeneity and Place-Based Rights

The importance attached to locality as a constitutive feature of indigeneity connects the latter concept to that of place as a basis for rights claims in general. On the view of advocates of place-based rights, a right to place is, in essence, a right *to remain* in place (Anguelovski and Martinez-Alier 2014, 173). Before the argument can be made justifying such a right, certain terms must be clarified, in particular, "place" and "homeland." A place is a "named locality" (Relph 1996). It is not simply a point on a spatial grid, nor an arbitrarily specified location. The name indicates a history and meaning to the particular locality. "Homeland" is a specific type of place. It is a "region (city or countryside) large enough to support a people's livelihood" (Tuan 1977, 149). So the concepts of place and homeland seem suited to connecting the two elements identified earlier that distinguish indigeneity—ancestral ties to land, and interest in sustaining the land so as to maintain peoples' livelihoods.

If indigeneity denotes a connection to a place that is the context for the ongoing existence of a people, then it is a homeland, and indigeneity is a type of attachment to homeland. Of course, it has its special characteristics. Primary among these is the sense of ancestral lineage that attaches to particular places. This is not necessarily true of other attachments to place, though they may have other features in common with indigenous claims to territory. This sense of ancestry gives indigenous peoples a "spiritual" connection to land in addition to the material connection of belonging to ecosystems and needing the land for survival. In this way, indigenous peoples have an advantage when it comes to the political struggle to retain or take back lands and territories that are in danger of loss or have already been lost. Indigenous peoples can, for this reason, often take a leading role as land and water "protectors" who are able to claim long-term historical and even prehistorical ties to the lands and waters that are subject to conflicts and struggles.

Nevertheless, traditionally indigenous peoples are not the only proponents of struggle over places. It is all those who need a connection to their local lands to maintain their livelihoods that can also claim place-based rights. This, of course, includes food producers—whether indigenous or not—who require control over lands or waters in order to maintain their subsistence economies. In addition, even peoples who inhabit distinctive urban places—neighborhoods—and live and work within their confines, may be said to be place-based in an analogous way to indigenous peoples. The generalization required here is what Bryan Norton calls "place-based environmental values" (Norton and Hannon 1997). These values derive from specific places and the needs and opportunities that they generate:

> The specification and explanation of sense-of-place values comes to this: for each community, provided its members are committed to preserving their community's sense of place, there are certain options—opportunities to engage in certain behaviors and authentic practices—that must not be lost if the community is to co-evolve with its place, adapting its practices to the cycles, as well as the capabilities and its limitations, of local environmental systems.
>
> *(Norton 2005, 432)*

This is as true of certain (though not all) rural and urban places, as it is of traditional indigenous lands.

Place-based values then are those that derive from the requirements of specific places, rather than from more general principles of justice or needs. It is this that makes indigeneity a feature of all place-based values: to be indigenous to a place is to know the values that sustain that place and its inhabitants. Yet, it remains the case that there is a gap between this sense of place and a right to place. What generates the right?

A right to place is claimed by those involved in ecological distribution conflicts—conflicts over environmental goods and services (Martinez-Alier 2002). For some, these conflicts involve a life-and-death struggle to retain control over the goods necessary for survival. For others, it is more a simple struggle to *remain in place* in order to retain a way of life that is valued and honored. This is a struggle that can occur in a number of different types of places, including urban ones, where, for example, working-class neighborhoods fight to retain control over their housing and services. The justification of the right, as with those instances in which the sustainability of rural ecosystems is threatened, lies in the attempt to preserve existing and sustainable neighborhoods and villages, as well as lands such as forests, plains, and farmlands. The maintenance of housing for the poor—provided this housing is suitable to a distinctive locale (perhaps as a result of the use of vernacular architecture or local food production)—could also be considered a type of place-based right.

Sometimes these different types of place-based livelihoods come together. In North America, there are Native American neighborhoods in urban settings (e.g., in Minneapolis, Chicago, and other cities), as well as in rural locales. Similarly, there are indigenous farming communities (e.g., the Onondaga Nation in New York) that claim the right to place so they can farm traditional foods; as a spokesperson states, "if we can grow our own food, we have sovereignty" (Maillard 2022). All these instances show that indigeneity is an instantiation of a broader conception of the relations of livelihood, place, and natural economy—a conception that encompasses myriad ways in which peoples possess an intrinsic attachment to homelands, and therefore can legitimately claim a right to the territory of which their homelands consist.

How are such claim-rights best understood? First and foremost, they are sovereignty claims—but of a type of sovereignty that is quite distinct from the early modern statist sovereignty that is assumed by most political theorists. It is a sovereignty that is not understood in "terms of assimilation, tradition, race, [and] nation," but as the "sovereignty [indigenous] ancestors and current elders have been speaking of in their expressions of locality" (Burkhart 2019, 92). The exact nature of such sovereignty claims—especially in relation to self-determination—will be specified in Chapters 5 and 6. But here we can focus initially on how the place-based nature of the claim-rights results in a certain understanding of rights to land and territory. There are two such possible ways they can be defined. On the one hand, they may be regarded as proprietary claims: claims to private or public property in land and resources. But the problem with regarding indigenous claims in this way is that indigenous land tenure systems often do not include either private or public property, since they lack markets or states as such. At most, they may be considered forms of communal property, with ownership held by a "commons"—for example, the people, community, or tribe.

On the other hand, indigenous claims to territory may be regarded as political: that is, they may be considered claims to self-determination, thereby justifying (indigenous) sovereignty over territories. Existing governments invariably regard indigenous claims in the first way, since property claims are usually, within modern property law, fungible claims, and also because regarding territorial claims as property rights (only) would not legitimate possible challenges to the sovereignty of states. Indigenous peoples, on the contrary, tend to view territorial claims as at least implicitly political, since only a more comprehensive (and usually non-fungible) claim would satisfy the need for rule over ancestral or other territories *in perpetuity*.

How to adjudicate these two interpretations of indigenous territorial claims? First, it would seem that viewing them as property rights cannot survive the understanding of indigenous land tenure systems as non-market institutions. Were indigenous peoples to want capital with which to maximize their position in markets, property rights would be the best way to do this, precisely because their fungibility would enable, under certain conditions, indigenous peoples to accumulate capital (e.g., by selling lands to mining companies). But this is not the point of indigenous territorial claims, at least to the extent that they are compatible with the core value of the ecological sustainability of the land. This means that the political interpretation that indigenous peoples themselves tend to adopt seems to be the right one. But what about the worry of states (and not only states) that such claims could produce challenges to state sovereignty that would eventuate in civil conflicts, to the detriment of all?

The concern about the increased potential for conflict as a result of indigenous territorial claims seems to be misplaced. That is because the meaning of self-determination for most indigenous groups does not justify statehood as a desideratum. States are the *problem*, not the solution for indigenous self-determination. The goal of indigenous self-determination is to "make states more 'inclusive and democratic'" (Karlsson 2006, 64). Of course, states could view *this* goal as threatening their exclusive state sovereignty. But this would be a misrepresentation of indigenous sovereignty claims. As Charles Mann writes, "When [North American] Native people talk about . . .'sovereignty'. . . it stands for a vision of Native societies as autonomous cultures . . . working as equal partners with nontribal government at every level" (Mann 2022, 46). As will be discussed in subsequent chapters, self-determination and sovereignty are neither incompatible nor focused solely on state power. Indigenous territorial claims are consistent with existing state sovereignty *as long as* those states are prepared to acknowledge and allow substate sovereignty over various matters (such as ecosystem maintenance).

Of course, this does not mean that there will not be conflicts or problems between indigenous peoples and states within which they are resident. At the same time, there are models for resolving these conflicts. Examples include new laws in Sierra Leone that require local consent for mining or other extractive industries (e.g., forestry) to operate, and that recognize the customary land rights of tribal/indigenous peoples (Peltier 2022). A proposed new constitution for Chile (ultimately voted down) recognized indigenous peoples as distinct nations and provided for their self-government within certain territories, including protections for their lands and resources (Lankes 2022). In addition, a number of states have recognized indigenous rights in some form, usually in their constitutions—for example, Bolivia, Canada, Ecuador, Finland, Mexico, New Zealand, Norway, Sweden, and the USA.

A constitutional framework within which indigenous rights are recognized in some fashion provides a means for potentially resolving conflicts between states and indigenous peoples. States should be viewed, not as neutral arbiters, nor as necessarily interested parties to land tenure and resource extraction regimes. Rather, they can be regarded as *potentially* compatible with indigenous rights claims to land and territory, given two conditions. First, there must be a vigorous political manifestation of indigenous peoples to: [1] take responsibility for the health and maintenance of ecosystems within which the indigenous reside, and [2] engage with states in order to ensure that there are not entrenched conflicts between state agencies and indigenous governing bodies. Second, there must be constitutional recognition of indigenous rights, so that, when conflicts do arise, states are not necessarily always arrayed *against* indigenous land and territorial claims.

Even if these conditions for resolving territorial disputes involving indigenous peoples are met, however, there will still be difficult and controversial cases. Some of these concern the status of existing, though largely ignored or violated, treaties between indigenous and nonindigenous inhabitants of territories. Jurists have means of adjudicating "conflicts of laws" and treaty violations, and perhaps these will be of help in resolving such disputes. In many cases, particularly but not exclusively in North America, indigenous peoples have expressed interest in reviving treaty provisions, giving them considerably more territory than they currently occupy. In Canada, for instance, there is a "Land Back" movement that insists on just this revival of treaty rights, and has taken various political actions to make it happen (Sen 2022).

Even when there are not treaty provisions that give purportedly just cause to indigenous claims to greater territories than currently held, however, the issue can still arise. This is because the real referent in indigenous claims is often, not treaties per se, but a memory and history of ancestral or traditional lands. The problem, of course, arises when indigenous peoples, who may possess, as either a "living" memory, or via historical documents, conceptions of ancestral/traditional territories, nevertheless live across wide and dispersed lands, even if they occupy discrete communities within them. Does this population dispersal invalidate the territorial claims? Or does the historical memory validate them? As Béteille writes, "The idea of indigenous peoples must have some basis in the territory inhabited by them in the past and the present. The problem arises when they become dispersed over large areas" (Béteille 1998, 190).

Complicating this picture is the idea, mentioned earlier, that indigenous peoples are often better stewards of intact ecosystems than either government land or forest agencies or local communities without traditions of sustainable land management. Does the seeming ability of indigenous peoples to sustainably manage lands make a compelling case for their reclaiming territories that they no longer inhabit (whether or not they have treaty-guaranteed rights to do so)? One example of this is the Apache people in the Southwestern USA. While the bulk (though not the entirety) of the people were moved out of the region in the late nineteenth century, some remain, and presumably others have memories about, the ancestral lands. Were they to make a bid for maintaining the lands, for instance, in the Gila Wilderness of New Mexico (part of the Apaches' traditional homeland), would this be a sufficient reason to grant them territorial rights? (Gwin 2023).

An additional complication is whether the lands are occupied (by others) or not. In the Apache example, the lands are largely unoccupied, which presumably could strengthen their claim. But occupation by settlers, even those who within living memory have expelled indigenous peoples, would complicate it. Yet, that too would not settle things. For instance, in the Palestinian case, admittedly one of the most complicated (and one where indigeneity is not often invoked), the presence of settlers ("occupiers"?) does not, at least in the view of expelled Palestinian Arab former inhabitants, decide anything about it.

There is no easy answer to this problem. Part of it will concern the applicability of (violated) treaties today. Beyond that, though, it will be a matter of negotiation between states and indigenous peoples, *as at least equal parties* to the negotiations. States do not deserve priority here, in part because *their* legitimacy depends on the extent to which they realize the self-determination of "their" peoples (including indigenous peoples). But in cases where treaties do not provide a reference point, there seem to be many possible considerations that could, under certain conditions apply. Which would be the more weighty ones? The aim of the current chapter is not to decide or settle these cases, only to point to them. This constitutes work for other philosophers to do, weighing competing considerations and deciding whether there are any overriding principles that might be brought to bear here to decide the hard cases.

The general principle should be clear, however. Indigenous peoples have rights over, and duties to, territories because that is partly what defines them as being indigenous. But this remains true only if they responsibly (i.e., sustainably) manage the territories for present and future generations. This duty to sustainably maintain ecosystems constitutes an essential basis both for assertions of indigenous identity and for their territorial claims. Conversely, failure to discharge this responsibility will invalidate such claims.

Both the food sovereignty and indigenous rights movements make claims to some version of sovereignty rights based on their connections to place, land, or territory. It is time to disentangle these different manifestations of the material underpinnings of political struggle over natural goods. At the same time, it remains important to remember that any such claims are contingent on the responsibilities of place-based communities, land/water protectors, and territorial guardians to maintain intact and in place these natural goods upon which depend the livelihoods of human inhabitants and the very existence of many other living beings. These aspects are brought together *theoretically* by the correlativity of rights and duties that inhere in all systems of "ethical life." They are also brought together *practically* through the espousal of sovereignty claims by peoples who know themselves to be inhabitants of particular ecosystems. These peoples—sometimes characterized explicitly as ecosystem peoples—will experience such rights and duties as those attached to their homelands, to the places and environments in which they must remain-in-place in order to pursue lives as distinct peoples.

The larger lesson here is that, in order to face and master the comprehensive environmental crisis that all living things are facing as a result of the industrialized extraction of materials from, and consequent degradation of, natural environments, peoples in general must find their place—their home—in the world, *remain in it*, and find the means to sustain the living systems that also reside there. It is evident that this, in and of itself, would constitute a radical reversal of historical trends toward human overpopulation, urbanization of these populations, and industrial production of necessities for them, especially food. Yet, as indicated in Chapter 1, there are plentiful indications that some version of this reversal will probably have to happen before the end of this century—due to resource exhaustion and the natural limits of intensive habitation under conditions of increasing scarcity.

The task for the present is to *protect* those who have found, or never left, their homelands, while they show the way to others for a different kind of life that is sustainable and sufficient. Inevitably, given the power of the market-state complexes geared to ever-intensifying industrialization, this task of protection will have to take on political dimensions, however much it may be rooted in prepolitical ways of life, habits, and beliefs. Indigenous peoples, food producers, and others must find their political voice, asserting rights of sovereignty in order to protect their ecosystems and the lives they seek to live within them.

The first theoretical part of this task is the clear specification of what needs protecting—lands, territories—and what protection means in this instance (e.g., property claims and/or territorial rights). What is needed is a way of disentangling the variety of ethicopolitical claims that might be made on the basis of homelands, natural economies, private or public properties, or sovereign territories. In philosophy, it is the latter, in contradistinction to various ideas of property rights, that has received the most attention. So it is with this that the theoretical inquiry into sovereignty claims must begin.

References

Alfred, Taiaiake. 1999. *Peace, Power, Righteousness.* Oxford: Oxford University Press.

Anaya, S. James. 1991. "Indigenous Rights Norms in Contemporary International Law." *Arizona Journal of International and Comparative Law* 8 (2).

Anguelovski, Isabelle, and Joan Martinez-Alier. 2014. "The 'Environmentalism of the Poor' Revisited: Territory and Place in Disconnected Glocal Struggles." *Ecological Economics* 102: 167–76.

Barsh, Russel Lawrence. 1994. "Indigenous Peoples in the 1990s: From Object to Subject of International Law?" *Harvard Human Rights Journal* 7: 33–86.

Baviskar, Amita. 2006. "The Politics of Being 'Indigenous.'" In *Indigeneity in India*, edited by Bengt G. Karlsson and Tanka B. Subba. London: Kegan Paul.

Béteille, André. 1998. "The Idea of Indigenous People." *Current Anthropology* 39 (2).

Buchanan, Allen. 2004. *Justice, Legitimacy, and Self-Determination: Moral Foundations for International Law.* Oxford: Oxford University Press.

Burkhart, Brian. 2019. *Indigenizing Philosophy Through the Land: A Trickster Methodology for Decolonizing Environmental Ethics, and Indigenous Futures.* East Lansing: Michigan State University Press.

Carneiro da Cunha, Manuel, and W. B. de Almeida. 2000. "Indigenous People, Traditional People, and Conservation in the Amazon." *Daedalus* 129 (2).

Corntassel, Jeff. 2008. "Toward Sustainable Self-Determination: Rethinking the Contemporary Indigenous-Rights Discourse." *Alternatives* 33: 105–32.

Daes, Erica-Irene A. 1996. "On the Concept of Indigenous People." *Working Paper for the U.N. Commission on Human Rights.*

Dahbour, Omar. 2003. *Illusion of the Peoples: A Critique of National Self-Determination.* Lanham, MD: Lexington Books.

Gregg, Benjamin. 2020. "The Indigenous Rights State." *Ratio Juris* 33 (1).

Guha, Ramachandra, and Joan Martinez-Alier. 2000. "The Environmentalism of the Poor." In *Varieties of Environmentalism: Essays North and South.* London: Routledge.

Gwin, Peter. 2023. "Wild West." *National Geographic*, June.

Hämäläinen, Pekka. 2022. *Indigenous Continent: The Epic Contest for North America.* New York: Liveright Publishing Corp.

Hart, H. L. A. 1983. *Essays in Jurisprudence and Philosophy.* Oxford: Oxford University Press.

Hegel, G. W. F. 1991. *Elements of the Philosophy of Right*, translated by H. B. Nisbet. Cambridge: Cambridge University Press.

Karlsson, Bengt G. 2006. "Anthropology and the 'Indigenous Slot': Claims to and Debates About Indigenous Peoples' Status in India." In *Indigeneity in India*, edited by Bengt G. Karlsson and Tanka B. Subba. London: Kegan Paul.
Karlsson, Bengt G., and Tanka B. Subba. 2006. "Introduction." In *Indigeneity in India*, edited by Bengt G. Karlsson and Tanka B. Subba. London: Kegan Paul.
Kingsbury, Benedict. 1996. "'Indigenous Peoples' as an International Legal Concept." In *Indigenous Peoples of Asia*, edited by R. H. Barnes, Andrew Gray, and Benedict Kingsbury. Association for Asian Studies Occasional Paper Series, No. 48. Michigan: The Association for Asian Studies, Inc.
Kingsbury, Benedict. 1998. " 'Indigenous Peoples' in International Law: A Constructivist Approach to the Asian Controversy." *American Journal of International Law* 92 (3).
Lankes, Ana. 2022. "Chileans to Vote on a New Constitution That May Transform Indigenous Rights." *New York Times*, September 3.
Maillard, Kevin Noble. 2022. "Native American Movement Grows." *New York Times*, August 31.
Mander, Jerry. 2006. "Epilogue: Summary and Final Comments." In *Paradigm Wars: Indigenous Peoples' Resistance to Globalization*, edited by Jerry Mander and Victoria Tauli-Corpuz. San Francisco, CA: Sierra Club Books.
Mann, Charles C. 2022. "We Are Here: North America's Indigenous Nations Are Reclaiming Their Sovereignty, Control of Their Land, Laws, and How They Live." *National Geographic*, September.
Martinez-Alier, Joan. 2002. *Environmentalism of the Poor: A Study of Ecological Conflicts and Valuation*. Cheltenham: Edward Elgar.
Niezen, Ronald. 2003. *The Origins of Indigenism: Human Rights and the Politics of Identity*. Berkeley: University of California Press.
Norton, Bryan G. 2005. *Sustainability: A Philosophy of Adaptive Ecosystem Management*. Chicago, IL: University of Chicago Press.
Norton, Bryan G., and B. Hannon. 1997. "Environmental Values: A Place-Based Theory." *Environmental Ethics* 19: 227–45.
Peltier, Elian. 2022. "Sierra Leone Laws Aim to Protect Environment." *New York Times*, August 10.
Relph, Edward. 1996. "Place." In *Companion Encyclopedia of Geography*, edited by I. Douglas, M. Robinson, and J. Williams. New York: Routledge.
Sen, Jai. 2022. " 'Land Back' Is More Than a Slogan for a Resurgent Indigenous Movement." *Globe and Mail* (Toronto).
Sengupta, Somini, Catrin Einhorn, and Manuela Andreoni. 2021. "A Global Plan to Conserve Nature Is Lacking a Vital Ingredient." *New York Times*, March 12.
Tauli-Corpuz, Victoria. 2006. "Our Right to Remain Separate and Distinct," and "The Prospect Ahead." In *Paradigm Wars: Indigenous Peoples' Resistance to Globalization*, edited by Jerry Mander and Victoria Tauli-Corpuz. San Francisco, CA: Sierra Club Books.
Tuan, Yi-Fu. 1977. *Space and Place: The Perspective of Experience*. Minneapolis: University of Minnesota Press.
Waldron, Jeremy. 2007. "Why Is Indigeneity Important?" In *Reparations: Interdisciplinary Inquiries*, edited by J. Miller and R. Kumar. Oxford: Oxford University Press.

PART II

Territorial Rights

3

THE NEW IMPORTANCE OF LAND AND TERRITORY

Place and Politics

The movements described in the first two chapters—movements dedicated to opposing extractivism and its environmental consequences—seem to operate outside of the usual avenues of political power and authority. This is both an advantage and a disadvantage of these movements. They are motivated by needs and demands emanating from the "lifeworld"—the felt experiences and urgent concerns of people for their livelihoods and the ecosystems that sustain them. This gives them a power and substance that "party politics"—including in some instances the politics of organized environmentalism—does not have. At the same time, without challenging directly the institutions of political authority, and the political theories that justify them, such movements will continue to act at a disadvantage in relation to extractivist agencies that can mobilize entrenched sources of wealth and power.

The argument of this book is that such movements require politicization and theorization in order to withstand, and to ultimately defeat, the forces acting to exploit and destroy life-sustaining ecosystems around the world. The first step in such a process is to consider the bases of such politicization—its foundation in the human experience of territoriality. This chapter and the next will not offer a general theory of territoriality; rather they will attempt to connect some basic concepts of, and ideas about, territoriality and kindred notions with aspects of political theory that are relevant to justifying anti-extractivist movements. Such concepts and ideas include those of *place*, *land*, *home(land)*, *property*, *commons*, and *territory* itself. In all cases, the emphasis will lie on the ways in which these notions either already are, or can be best understood to be, political. Such

DOI: 10.4324/9781003568155-6

understandings will, in the end, yield an *ecosystem-based theory of territorial rights*. Such a theory will be shown to be the best interpretation of the fundamental idea of territorial rights as such. It will accordingly constitute the basis for sovereignty claims that, as Chapters 5 and 6 will show, can challenge the entrenched power of extractivism in states and other political agencies.

The first step in this argument is to understand the enduring, and newly revived, importance of territory for political power, and human life generally. While this idea goes against dominant strands in political philosophy of the last generation or so, it does accord with the increasing value of land and territory in contemporary economic activity, as well as its increasing importance in recent political conflicts. This phenomenon has been recognized by geographers and other social scientists, but it has yet to make a major impact on political theory. As the present chapter seeks to show, however, it needs to be taken into account when justifications of political claims are made.

A good place to begin is by considering the relation between political life and geographical places, which is the subject of John Agnew's classic work, *Place and Politics* (Agnew 1987). Agnew makes a number of important points about the relation of place to political theory, of which two are especially relevant for our consideration of territoriality. While Agnew's diagnosis of the "devaluation of place" in political theory is important, what is more crucial for our analysis is the ways in which he disambiguates the concept of place in terms of its various meanings (Agnew 1987, 62–79). Ostensibly, place and "locale," in the sense of a particular point on a spatial grid, are largely coextensive. But Agnew argues that locale is only one of three different dimensions of place. While locales, as the "settings in which social relations are constituted," are important, the dimensions of "location" and "sense of place" are equally pertinent (Agnew 1987, 28). To begin with, sense of place is the "local 'structure of feeling,'" one that "can be *projected* onto the region or a 'nation.'" As he puts it, "The sense of place need not be restricted to the scale of the locality." Similarly, location designates the interaction between locale and other, non-local dimensions of social life; it is the "geographical area encompassing the settings for social interaction as defined by social and economic processes operating at a wider scale."

The significance of these definitional specifications for political action is twofold. Both concern the relations of sovereign territorial states with the regions or places that constitute, but are not coextensive with, them. First, it is important to note that territorial states, in order to maintain their sovereignty, both internally and externally, require a high degree of integration of the different places that comprise their territories. This is something, as Agnew notes, that is not easy to achieve (or permanent when achieved):

> The central state . . . *depends* upon attracting support and legitimacy from large sections of the territory over which it holds sway . . . . [This is because]

> the successful and continuing integration of the modern territorial state requires a considerable degree of popular consent and participation in political life.
>
> *(Agnew 1987, 39)*

Second, the relation between states and substate regions or places within them is crucial, especially under conditions of inequality between these places: "under conditions of economic and political inequality between places, a hegemony can only be maintained by pursuing policies with geographical consequences and place-biased policies that favor those people and places that favor the hegemony." Particularly in the case of large territorial states, comprising many different places and regions (e.g., Brazil, China, India, Russia, and the USA), state power is predicated on the mobilization of some places, in order to dominate, and in some cases exploit, other places within the state's territory. All this is to say that regarding sovereign territorial states as "black boxes," containers of uniform places and peoples, is a fundamental mistake, at least in terms of explaining political behavior. As we will see, it is also a mistaken approach to articulating legitimate political rights and duties.

What this will require, aside from some detailed consideration of territoriality and its connection to political action, is a deeper understanding of how *places* are constituted from *spaces* and how places are connected to peoples. But first, it is important to consider whether the predominant, cosmopolitan tendency in the last generation of political theorists is able to encompass some consideration of place-based rights (such as territorial rights) into the theory. It will turn out that there are probably insuperable conceptual difficulties to doing this, and that acknowledging the importance of place, land, and territory for political theory will require fundamental changes in how we think about collective rights and political legitimacy.

Against Cosmopolitanism

As previously noted, the recent revival of interest in territoriality is occurring in part as a reaction to the dominance of cosmopolitan ideas in recent political philosophy. The discourses of cosmopolitan democracy, human rights, global governance, international legal reform, and global redistributive justice that have constituted much of post-Cold War political theory have left a massive deficit when it comes to determining the sovereignty, membership, boundaries, and rights of actually existing political communities. That these communities have not disappeared as a result of the development of global financial, mercantile, diplomatic, and military institutions is an affront to cosmopolitan hopes. But it is nevertheless the case that, for most of the world, politics remains local, not global, and the ability of peoples to control their destinies is very much contingent on the nature and extent of their particular communities.

Regardless of this, a few cosmopolitan theorists have addressed the issue of boundaries and territoriality, in an attempt to foreclose a more widespread reaction against the seemingly obvious inability of cosmopolitanism to speak to the actual conditions of political life for much of the planet. Can there be a viable cosmopolitan theory of territoriality? The first thing to determine in answering this question is what such a theory would have to do to be cosmopolitan. If there were a theory of territoriality that accorded rights to peoples or other entities to claim and protect particular territories, it, in order to be cosmopolitan, would have to also be consistent with some version of the principles of cosmopolitan or global justice.

Obviously, what constitutes such principles has been the subject of much controversy. Nevertheless, a couple of points can be made about it. First, no ethical principles, such as one of territorial rights, must override the right of global laws and institutions to enforce planet-wide principles of distributive justice, human rights, and so forth. Any territorial right, in short, must be conditional on its compatibility with cosmopolitan rights. Second, such rights are the result of theories and judgments that are based on the equal claim of all individual human beings to the same life prospects. This is the core of the cosmopolitan vision—and it is based on a thoroughgoing individualism that accords identical treatment to all persons globally.

It might seem that, with such foundational principles, it would be hard to justify the right of a people, state, or community to territory. Furthermore, it might seem as if cosmopolitans' fundamental political commitment must be, not to a particular community, but to humanity as a whole. An entailment of this view, and it is a common one among cosmopolitans, is that boundaries and territorial claims are conditional, contingent, and revisable in accordance with the principles just enunciated. However, there have been a handful of theorists who still contend that a cosmopolitan right to territory is a possibility. Two points must be substantiated for this to be the case. First, there must be some way of arguing for a limited, or particular, right to territory that does not easily devolve into a case for world government, a borderless world, and so on. Second, there must be a mechanism for determining actual boundaries and the legitimacy of territorial claims.

One might think that a cosmopolitan view necessarily implies a humanity-wide political community as the only legitimate one. Yet, Arash Abizadeh argues that this need not be the case. Despite his contention that there are no *inherent* limits or boundaries to democratically constituted political communities, he argues that this does not mean that there can be only one such community in the world. From his point of view, this implication would be just as arbitrary as claims to specific territories. Whatever the ultimate boundaries of territories, the question is whether they have been adequately legitimated. From Abizadeh's point of view, there are only two possible processes of legitimation—a "pre-political," nationalistic one, or a "political," democratic one. In the first case, there is a conceptual problem with basing legitimacy on pre-political agencies or subjectivities, such as that of national groups: "Any principle identifying some factual aspect as a ground of legitimacy presupposes a normative principle explaining *why* that ground has the capacity to legitimize political power" (Abizadeh 2012, 879). The idea that there needs to be a "prepolitically constituted corporate agent, a collective democratic subject bounded prior to the exercise of political power" is consequently mistaken; such an agent must itself be "politically" legitimated first. So we are left with the second case, a political process of democratic legitimation to determine territorial boundaries.

But the idea that such boundaries can be democratically determined is a problematic one, as we will examine in more detail later on. For now, we need simply to understand how a cosmopolitan theorist who views political community from a "one-world" perspective might argue for a democratic determination of territory. Such a determination is to be understood in "intersubjective, procedural terms," and is located, "not in a prepolitically constituted collective entity, but in the regulative standards implicit to democratic procedures and processes that make collective self-rule possible." Given that there is no prepolitical identity with which to limit the community, and that democratic procedures themselves have only conceptual, not geographical, constraints, this means that the "demos is *in principle* unbounded" (Abizadeh 2012, 880). Yet, this does not mean that the "only legitimate democratic polity comprises all humanity. Rather, it implies that particular boundaries can and must be legitimized as the outcome of democratic procedures that include those whom the boundary picks out as outsiders." So, Abizadeh seems to mean that for a territory to be determined in accordance with cosmopolitan justice, those excluded from the territory must agree to their own exclusion.

How this would not eventuate in a territory as being all-inclusive—for who would voluntarily exclude themselves from membership and rights in a community?—is not made clear. Yet, the idea that there could be a cosmopolitan justification for territorial rights is also held by Lea Ypi, who argues for a more institutional embodiment of cosmopolitan right. This embodiment she refers to as a "global authority," and it is the willingness to accept such an authority—which is itself presumably committed to the realization of cosmopolitan norms—that accords legitimacy to any particular state's claim to territory: "States can continue to exercise territorial rights if and only if their citizens are also politically committed to the establishment of a global authority realizing an all-inclusive principle of right" (Ypi 2012, 293).

One way to understand this idea is just that (only) states that accept and abide by international law are legitimate and therefore have the authority to make territorial claims. But what if these claims are in conflict with those of another state that also abides by global norms? The adjudication of territorial claims by acceptance of international legal doctrines would seem to be underdetermined, inasmuch as it does not have a means of settling disputes between equally law-abiding states. So the idea of a "global authority" must be taken more literally—it seems to require an actual political institution that would adjudicate rival claims. Without this, there does not seem to be any way to settle disputes when there is no principled disagreement, but simply overlapping claims to territory. As Ypi concludes, we must wait for the appropriate "authority" to adjudicate disputes—there is no other principled way to proceed: "The answer to the question of why states are entitled to one particular area of geographical space cannot be conclusively provided . . . until a political authority realizing the universal principle of right is founded" (Ypi 2012, 303).

Aside from postponing an account of territoriality into the (indefinite) future, such a view is problematic for two other reasons, one an entailment of the cosmopolitan view, the other a conceptual limitation of it. First of all, if a right to territory is legitimated solely by its consistency with a general conception of justice (e.g., a cosmopolitan one involving a bundle of notions of human rights, just distribution, etc.), then any political agency that is so consistent would be legitimately entitled to the territory. Not only does this underdetermine often highly contested claims, but it also does not exclude agencies that may be committed to cosmopolitan justice but are not politically legitimate. The obvious example is imperial or hegemonic states, ones that may endorse, for example, a "human-rights agenda," but are otherwise pursuing their own imperial interests. Cara Nine has put the point this way:

> If agents must only . . . demonstrate the capacity to meet minimal standards of justice, then any authority with the capacity for a minimally just regime could claim to be a potential territorial sovereign. On this criterion alone, 'benevolent' colonization . . . would be justified.
>
> *(Nine 2012, 50)*

In other words, justice does not confer legitimacy—these are two different considerations, and territoriality needs to be determined legitimately, not in accordance with extraneous principles of distributive justice (Moore 2015, 99–100).

The second reason that cosmopolitanism does not yield a theory of territory, however, is more fundamental. Any theory of territoriality must have a way of differentiating between particular territories, or more generally, places. While all such territories or places will be locations of human habitation, this fact alone means little in terms of which peoples legitimately warrant their attachments to which places. The problem with cosmopolitanism is not only that it has no conception of place (and therefore, of territoriality) but also that it *cannot* have one because of its basis, instead, in a conception of *space*. The individualism of cosmopolitan thought is linked to a matrix of abstract space, of which the planet is but one manifestation. There is in principle no difference between distinct points on the spatial grid; the only matter of concern is the persons who may or may not inhabit such points. But points are not places, and it is connections and attachments of peoples to such places that generate both the problem of territorial rights and, perhaps, its solution. This conceptualization of human-environmental relations as abstractly spatial is what makes cosmopolitanism, in Daniel Deudney's words, "weak and relatively ineffectual as a source of political identity and community" since it is "incapable of evoking or sustaining a robust [sense] of place" (Deudney 1996, 130).

From Land to Homelands

What cosmopolitanism lacks is an understanding of the *materiality* of territory. Territory consists primarily, not of moveable or fungible aspects of social life, but of immoveable and irreplaceable domains of planetary space. Territories, in other words, are not only specific places but also bounded spaces, or specific lands. Land, from the perspective of much twentieth-century political theory, seems to embody a misplaced concreteness. As an economic factor, a cultural constant, and finally, a political good, land appears to be inessential, as compared to other aspects of economic, cultural, and political life. The market, the nation, and the state seem to "hover above" land, alighting only to instantiate their already constituted bundles of wealth, identity, and power in a somewhat arbitrary location.

In much of late modern political philosophy, land had apparently lost the overriding importance given to it by such early modern thinkers as Quesnay, Herder, and Locke, who accorded centrality to the interaction of people with the lands upon which they depended for sustenance, meaning, and legitimacy. Yet, in a startling reversal, land has started to become a twenty-first-century obsession, producing "land grabs" and displacing prior interests in financial capital flows, cultural fluidity and "hybridity," and transnational or global political institutions. For two centuries, the material location and constitution of communities seemed to be inessential or incidental in comparison to more important factors such as economic productivity, cultural integrity, and political history. Yet, today, it is more and more recognized that land provides the ineliminable context for human communities, including their material sustenance and spiritual meaning. This is why territoriality today is to be understood in relation to land—because land itself has become increasingly valuable and therefore contested.

There are several reasons for this. First, there is the increasing demographic pressure on land use, with an exploding global population clearly having crossed the threshold of sustainability. Second, the overexploitation of cropland has resulted in soil exhaustion and declining crop yields, motivating the search for new land purchases. Third, an increasing use of cropland for "biofuels," rather than food, has led to competition between food producers and energy companies for land. Fourth, the search for nonfungible wealth, as a result of economic crises (that have led to the collapse of capital markets), has led to an increasing "financialization" of land—the purchase of land as a relatively secure investment.

The demographic crunch, especially in poor regions of the Global South (e.g., Bengal and Punjab in South Asia, Nigeria and Ethiopia in Africa), has resulted in tremendous pressure on farmland, forests, and ecosystems that, in prior centuries, were able to sustain significant human and nonhuman populations. While population pressures obviously correlate with consumption levels of populations, aggregate studies have suggested that planetary carrying capacity has already been exceeded, though to what extent is a subject of controversy (Dovers and Butler 2015). In any case, land use (and value) is obviously affected by increasing space needs for human habitation, especially in built environments requiring substantial development of artificial infrastructure (Purdy 2019, 83–84).

A significant amount of land globally is already devoted to agricultural usage, often with "hard-till" plantings and intensive use of chemical fertilizers that stress and degrade soil health. The result can be exhaustion of agricultural lands, decline in crop yields, and a search for new acreage that can support such intensive exploitation. The actual (and much greater potential) declines in food production from this overexploitation are resulting in increasing pressure to find and cultivate new lands (Pimentel and Burgess 2013).

Land is also becoming a source of energy as well as food, due to the increasing expense and cost of fossil fuel production. As Alf Hornborg has shown, the traditional use of land managed to produce *both* food and energy (in the guise of animal fodder); but this dual usage was displaced by the advent of fossil fuels in the early nineteenth century:

> Access to fossil fuels . . . in the early nineteenth century obscured our fundamental dependence on land. When acreages for horse fodder and food for workers could largely be replaced with coal mines and oil wells, present-day land became . . . a substitutable factor of production.
>
> *(Hornborg 2012, 19–20)*

But today, as coal mines and oil wells disappear, or get prohibitively expensive to exploit, land is once again valuable for both food *and* energy. From biofuel cultivation to solar panel arrays, land stands to become scarce, expensive, and desirable for a multitude of uses:

> As we are currently contemplating that peak oil and climate change may prompt us to turn to biofuels, we are in fact imagining a future where land

> requirements and energy requirements once again coincide . . . . [We may] once again see competition over scarce land for food, fodder, fibres, and fuel.
>
> *(Hornborg 2012, 19)*

Beyond the pressures of population growth, food production, and energy use, the search for new areas of financial growth, in the period following the 2008 financial crisis, has turned increasingly to land as a promising avenue of capital investment. This has generated the "financialization" of land to an increasing degree, where land is regarded as a nonfungible and therefore more stable investment opportunity. Agricultural land, in particular, has been sought by wealthy companies and countries looking to secure sources of wealth (both food and energy) for their growing populations. In this way, all of the factors valorizing land in a new way come together to generate financial pressure and competition over increasingly sought-after (and expensive) land resources (Fairbairn 2020).

Land and place have therefore become coextensive terms, at least inasmuch as the concern is with the material conditions for generating specific loyalties to local political institutions. Such institutions have always provided the basis for political legitimacy; yet, contemporary sensibilities have suggested otherwise—that legitimacy could be derived from abstract, non-place-oriented considerations. The site of human communities on specific landed domains, rather than across multiple decentered spaces, suggests that land is again playing the role that it had for premodern societies. Yet, it is the subjective, "spiritual" conditions for political life that also must be considered in understanding territoriality in its full dimensions—especially the human attachments to land embodied in conceptions of home and homeland.

Concern for the preservation of environments has historically often been a matter of *local* attachments. When this fact is ignored, environmentalism becomes the business of policymakers and lawyers, but not of people(s) with important attachments to places, lands, and territories. Love of homeland [*Heimat*] has been a frequent feature of human history, but not concern for the planet. The environmental historian Joachim Radkau puts it this way:

> [The original] nature protection movement [had] a foundation in love for *Heimat* [home] and in the attachment to a familiar image of one's native land. For it is only such a guiding image that can be infused with popular support and desire—a threshold value for emissions that is negotiated among experts is not the kind of goal that inspires people.
>
> *(Radkau 2008, 236)*

It is sometimes thought that attachments to localities and homelands is a function of nationalistic sentiments. But this is a mistake: it is more probable that nationalist loyalties are themselves often grounded in local attachments, than vice versa. Of course, as I have argued at length elsewhere, nationalistic sentiments, both because of their justification by ethnonational narratives, and of their obsessive focus on attaining state power, are corrupted expressions of these attachments (Dahbour 2003, 91–114). This is partly the result of their conflation of two quite different sorts of attachment—one based on contiguity (place) and the other on consanguinity (kinship). In fact, these two are further conflated, as Daniel Deudney has argued, with a third element of patriotism and allegiance to states. As Deudney has put it (following the geographer Yi-Fu Tuan), much nationalist rhetoric indulges in "topophilic" sentiments—evocations of places and conceptions of homelands—while actually attempting to turn such attachments to the service of movements to create nation-states (Deudney 1996, 131–32).

Deudney furthermore points to contemporary environmental movements as "challenging the pseudo-primordial claims of nations and nationalisms" by linking such attachments to place with the science of ecology. While he may be right that this is a promising new development, he misses the point that, prior to modern environmentalism (though perhaps acting as a source for it), attachment to place (i.e., homelands) went very far back in human history, as Radkau has argued. The attachment, even without the overlay of contemporary environmentalism, is a basic motivation underlying territorial claims that is distinct from the state-oriented politics of nationalist movements.

More fundamentally, however, local attachments, first to a land, and then, more specifically, to a place—a *named* locality—often generate powerful sentiments of *belonging* to that place (and land) (Relph 1996, 906). These sentiments have been understood as designating attachments to a home or homeland—the German term, *Heimat*, captures both of these connotations. Homelands can be inherited—from families or communities or cultures—or they can, in some cases, be rediscovered or revived (Scruton 2012, 237). The love of, or attachment to, homes and homelands—*oikophilia*—seems to have a strong affinity with two aspects of human life. On the one hand, it is often the context for strong expressions of sharing and belonging with others in settled communities. These expressions often become ones of self-sacrifice—whether of work, time, or even, at the extreme limit, life—for others in the homeland (Scruton 2012, 239).

On the other hand, homelands are the most powerful contexts within which humans connect with, and become attached to, the natural world in its various manifestations (Scruton 2012, 231). Homelands, as various naturalists and "nature writers" have emphasized, are the concrete, specific "environments" within which persons live, and for which they care—not an abstract concept of the "global" environment (Kingsnorth 2017, 72, 80). When these homelands are lost, neglected, or rejected for the dubious pleasures of a "consumer society," the natural worlds within which they exist are often similarly degraded or destroyed (Scruton 2012, 241–42, 246). And when this happens, often in the face of continuing losses of nonhuman life and intact ecosystems, it is only the building of "refuges," constructed out of the remnants of our (former) homes, that holds the promise of recovery and renewal (Kingsnorth 2017, 146).

It is important to note that this attachment to localities, understood as homes or homelands, is not an exclusively "sentimental" one. Such attachments accord with the results of recent research into the nature of "socioecological systems," and the norms or values that allow such systems to survive. Of course, from the perspective of political ecology, all human communities are socioecological systems, of one kind or another. But the fact is that, perhaps uniquely among animal species, humans do not share one design for adaptation to natural environments. Variability in adaptation is a special feature of human interactions with their environments, since, as Joan Martinez-Alier puts it, "such adaptation depends not on 'nature,' but on economics, politics and culture" (Martinez-Alier 2002, 26).

Nevertheless, there have been commonalities across most human cultures in terms of the ways in which natural goods have been utilized to provide ongoing bases for the social reproduction of human beings. Above all, human cultures have generally regarded themselves as being embedded in nonhuman contexts:

> With few exceptions, including the Western industrial societies of the last four hundred years or so, human societies have generally regarded themselves as *part of nature* and not separate from it . . . . [However, there] is no single universally accepted way of formulating the linkage between social systems and natural systems.
>
> *(Berkes and Folke 1998, 9, italics in original)*

This self-understanding of human cultures that exist in a relatively direct and explicit relation to natural systems accords with the general thesis of socioecological systems theories that "[a] society's survival is ultimately dependent on the finite capacity of ecosystems to support it with essential resources and ecological services" (Folke, Berkes, and Colding 1998, 433).

There is, though, at least one specific feature of successful socioecological systems throughout human history that can be clearly distinguished—the *local* scale of most such systems in using, but also sustaining, the necessary resources for social reproduction. In contrast to industrial systems, most others "do not undermine their existence by degrading their ecological life-support system, thereby losing ecological and institutional resilience . . . . *The local community has become a part of the dynamics of its surrounding ecosystem*" (Holling, Berkes, and Folke 1998, 356). *Resilience*—the ability of systems to recover from environmental setbacks and shortages through planned redundancies in resource maintenance—is a key value for socioecological systems other than industrial ones. The locality-based scale of such systems includes "feedback loops" that provide information leading to restrictions on unsustainable resource use: "local social systems of rights and responsibilities develop for any resource deemed important for a community . . . rules arise and evolve according to local needs" (Berkes and Folke 1998, 7). Absent this local involvement in and attachment to particular natural systems, it becomes difficult to regulate and maintain sustainable resource use: "[If] communities are not involved in the active management of their natural resources, they will use resources destructively" (Agrawal 2000, 39). Local attachments then serve an essential role in identifying homelands based on particular ecosystems that provide the necessary context for a sustainable and resilient scale of natural resource use.

We have been considering a series of concepts—place, land, home(land)—that, though they have been virtually forgotten by political theorists for some time, are now essential to understanding the historical moment in which we find ourselves. They designate, in various and increasingly politicized ways, the material, geographical, and ecological context in which aspects of the contemporary environmental crisis are playing out. While places are the basic building blocks of geographical knowledge, they have acquired renewed importance as human responses to ecological degradation grapple with the differential effects of anthropogenic environmental change. Aside from changes in atmospheric conditions affecting climates, land degradation is the most extensive of these changes, with multiplier effects on biological populations and human subsistence itself. With the demographic upheavals resulting from these changes accelerating, maintaining homelands—or in cases of ecological refugee populations, finding new homelands—has become an increasingly urgent matter for millions of people.

Yet, political theory—especially the quasi-contractarian theories that justify political liberalism—lack concepts beyond the idea of property ownership that can specify the justice or injustice of specific claims to places, lands, and territories. In fact, the very idea of territory is often confused with that of property—at least in terms of what justifies rights to or over it. Without getting clearer about the difference between proprietary and territorial claims and rights, it will be difficult to understand the justice or injustice of particular distributions of natural goods and ecosystem services.

Property Ownership or Territorial Rights?

For political liberals, as for early modern contractarians, the main political question was always about the proper relation of states to *individual persons*. Such persons were often thought to be property owners—as in Locke's legislative authority comprising those possessing private property or in Rawls' "property-owning democracy." This meant that, if states were to exercise their sovereignty over particular territories, this could be justified best if it was understood either as founded upon a property right, or as itself a property right of a special kind. In either case, it is notable that territorial rights are legitimated on the basis of individual rights of some sort. There is no role in this theory for communal rights to territory, whether of agrarian communities, indigenous peoples, or other collective associations.

This is why the beginning point for much recent thinking about territorial rights has been whether such a right is or is not similar to a right to property, as that has been legitimated by political philosophers. For instance, some contemporary philosophers advocate an explicitly *proprietary theory* that has been derived to some extent from Locke's theory of property rights. Both John Simmons and Cara Nine, in different ways, espouse this sort of theory of territorial rights (though Nine, in her recent work, has moved away from an explicitly proprietary approach; see Nine 2022, and the discussion of this in Chapter 4).

Nine makes a distinction between a theory of this type that is individualistic and one that is collectivistic (Nine 2008, 148). The first attempts to derive a territorial right directly from an individual's (prior) right to their own property; the second develops a territorial right from the value of a state's jurisdiction over territory, in a sense comparable to the value of private property (for individuals). Simmons' conception of territorial rights is an example of the first view, Nine's of the second. Both rely on the basic idea of the Lockean justification of private property and the state. Private property for Locke is justified by actions of individuals prior to the legitimation of government; it is the result of the productive use and improvement of a portion of the natural world by individual persons. The state is then justified as an efficient mechanism for the protection of the land and other resources used and improved by productive individuals. These individuals consensually agree to the limited authority of the state to act in this manner.

What is not clear from the Lockean account is *where* the state can exercise its authority. While it is clear that it can do so over the private property of those individuals consenting, the agglomeration of this property does not necessarily comprise a governable territory. This is the first problem with a proprietary account. As Simmons notes,

> For Locke's account to end up justifying anything remotely like the territorial claims of actual modern states, he needs to explain why they should end up with legitimate *continuous* boundaries, not just how they can end up with legitimate jurisdiction over various chunks of land.
>
> *(Simmons 2001, 313–14).*

Simmons thinks that Locke makes two implicit assumptions that allow for territorial rule—a ban on "interior dissenters" (i.e., "landowning persons who decline membership but whose holdings are surrounded by those of consenting members") and a right to ownership of "unowned interior land" by consenting members of the surrounding territory. Now Simmons is somewhat unimpressed by these assumptions, but regards them as providing the rudiments of a territorial right, albeit one that is without the considerable force that modern states would usually like to employ. From his point of view, however, states on the Lockean-consensual view lack full legitimacy anyway, since consent theory implies the ability of individuals to opt out of accepting political authority. For Simmons, the foremost contemporary exponent of Lockean political theory, its quasi-anarchistic implications are actually an advantage of the theory, because of his more general view of the purpose of political theory overall. As he writes, "Theories of legitimacy and justice should, in my view, have as their contents genuine moral ideals, uncompromised by allowing moral justification to be held hostage to a rationalization of our own familiar or preferred political arrangements" (Simmons 2001, 315).

But this view also leaves the territorial right of states on very shaky ground, to the extent that it requires a basis in the property rights of consenting persons. This is why Nine, who likes certain aspects of the Lockean theory, nevertheless insists that there is a better, collectivist, version that can provide a more robust conception of territorial rights. From her point of view, if the Lockean theory is taken over directly in the way that Simmons attempts, it will make territorial rights a direct right of individual property owners. But this is to hold it hostage to individual dissenters who can legitimately take out "chunks," to use Simmons' term, of states within the borders of which they live, but whose authority they refuse to recognize (Nine 2008, 152–53; see also Steiner 1996, 139–48). Such dissenters can simply, and legitimately, exit, taking their "chunks" (i.e., lands) with them as they go.

Nine writes that this makes a mockery of the very idea of territorial rights, whatever other legitimacy the idea may have. Yet, she maintains that, properly modified, a theory of property rights can be extended to become a theory of territorial rights. The way to do this is by focusing not on the consent-based theory of government but on the original conception of property. In this conception, the property right is given by the productive activity of the individual who uses and improves the land that they thereby are entitled to claim as their own. Nine believes that a similar argument can be made for the productive activity of the state in relation to land use.

The general form of the argument is given in her book as follows:

> According to a Lockean theory, in order to acquire a right over a good, (i) an agent must be capable of changing the land thereby creating a relationship with it, and (ii) this relationship must be morally valuable—established through the fundamental liberal concern for basic individual needs expressed in terms of principles of desert, efficiency, and autonomy.
>
> *(Nine 2008, 82)*

The state, on her view, is able to do this, and consequently can claim a right to the territory on which it does so. The "relationship" established is one of the "efficient use of the land"; the "moral value" accrued is that of the meeting of basic human needs (food, shelter, etc.) (Nine 2008, 161–62). This right, according to Nine, is not a "maximizing argument"—either in the sense that it is indexed to productivity—nor that it could justify the appropriation of less efficiently used lands. A simple use for meeting human needs will do: "One does have to be using the land in a manner that tends to preserve the human species and in a manner that more efficiently promotes that end *than if the land remained in the commons*" (Nine 2008, 162; italics added).

Whether the Lockean theory is given an individualistic or a collectivistic interpretation, it relies fundamentally on the idea that territoriality, as a relation between peoples and territory, is the same as property, the relation between persons and resources. This relation is one of ownership: persons own property (under certain conditions), states own territory (under similar conditions). Is this the right way to understand this relation? I will bracket for the moment the question as to whether a Lockean conception of private property is the right way to understand the relation of *persons* to the environment—though considerable skepticism should be registered about that as well. But assuming that the property theory is viable, does it apply to territory as well, in some fashion?

There are three problems with the specific notion that territoriality can be understood in this way—a problem about occupancy of territory, a problem with the state as owner-occupant of territory, and a problem with the idea of use and improvement. First, in terms of occupancy, the theory seems to rely either on a right of *first* occupancy or a right of *present* occupancy. If it is the latter, the theory devolves into the view that acquisition confers rights, however, the territory is acquired (e.g., by force). This contradicts the very idea of rights-based claims to territory, since any new acquisition would invalidate previous claims, whether rights-based or otherwise. But what of the idea of first occupancy? Especially, since the maximization of the value of, or goods within, a territory is not an explicit consideration, on Nine's interpretation, it would seem that the first state to occupy territory gets the right to it in perpetuity, as long as it continues to "use and improve" in some fashion. But, since the first occupants have, almost everywhere, been displaced at some point historically, the right becomes mute. Only if the right is given a rigorously historical interpretation, something that Lockeans have not done, would a first-occupancy conception remain consistent (cf. Nozick 1974). But then it would also cease to have much applicability to a world in which environmental and demographic change are continual.

The second problem concerns the state, as quasi-owner of the territory. As Margaret Moore points out, this is a fundamentally conservative notion, inasmuch as it accords with older monarchical and dynastic ideas (in Europe, but also elsewhere) of the ruler as owning their domains. As she puts it, "If we think of territory as the property of the state or the sovereign, there are no conceptual resources to consider the wishes and well-being of the people who live on the territory" (Moore 2015, 24). Perhaps an expanded conception of the state, that does not separate it inherently from society and its inhabitants, can assuage a part of this concern. But the larger problem concerns the relation of ownership as the proper conceptualization of people-territory relations. Owners of property—and presumably rulers of territory—have rights over their owned (or ruled) possessions, and rights versus other owners or non-owners. Though property ownership is not generally regarded as allowing the use of property to *harm* other persons, this is a rather minimal restriction on use (Honoré 1961, 123). Nothing is implied about better or worse uses themselves—this is the purview of owners to determine. No other theories of state authority—communitarian, republican, utilitarian, idealist, or otherwise—regard the relation of the state to society as one of ownership and use of the latter by the former. At best, such a view is, as Moore suggests, fundamentally conservative: it does not seem to allow for a more emancipatory conception of state power.

Third, consider the idea that territory is owned in proportion to its use and improvement (i.e., the creation of value). While it may be the case that a strictly maximizing role for the state—in the sense of the maximization of economic value-added—is not mandated by the proprietary theory of territory, it is certainly not prohibited by it. Furthermore, there may be an "elective affinity" between the two concepts, as there also undoubtedly is with the Lockean conception of property. But more importantly, *who* is to determine the appropriate use, improvement, and maximization of territory, such that a right to that territory results? There are no conceptual barriers to regarding the state as determining what it should be doing with "its" territory; criticisms being limited to states that "do nothing" with it. Perhaps, states could also be criticized for wasting land and resources—though even the definition of waste is variable enough so that states could be exempt from much practical criticism on this basis. What is to prevent the systematic exploitation of nonrenewable resources as an imperative granting a state the right to territory (rather than voiding its right, as some might argue should be the case)? There just does not seem to be any *systematic* way to avoid (economic) value- and wealth-maximizing policies being pursued by states as the surest way to legitimate and perpetuate their right to territory on the proprietary account. This theory also seems to lack any way to legitimize a less exploitative and more sustainable approach to territoriality—in fact, such an approach may be prima facie *il*legitimate.

The overall problem with the proprietary theory of territory, then, is that territory cannot be unproblematically connected to the ways in which individuals claim, acquire, and use natural goods as private property. At best, ownership of territory might seem to be justified as a kind of *common* property right. But that is to treat it very differently from how ownership of places and spaces is regarded when it is viewed as a matter of individual entitlement. Common property regimes are just that—forms of governance (*regimes*) with political aspects that suggest the need for a very different form of justification. As we will see in Chapter 4, establishing a right to territory is an irreducibly *political* matter at least as much as it is also a function of ecological considerations. But, it is important, before we consider how such a right is to be understood, to be clear about to what a communal right to territory applies. It is, of course, land—and its legitimate uses—that is paramount here. Rights to territory, in other words, function as solutions to conflicts over land and its uses.

Natural Capital and Common Property Regimes

Land exists prior to human uses of it; in this sense, land is a *natural kind* (Kolers 2012, 277). But, to the extent that it comes to play a role in human life as a "natural resource," it is an *intentional kind*, linked to human purposes through its uses. It is *not*, however, a feature of a political economy in which value is produced through the employment of human labor (power). Land constitutes value even prior to its transformation through labor in various ways. It is a part of the "natural capital" that is distinct from the capital accumulated within economic production and exchange relations. In that sense, it is a part of the "natural economy" that provides the often unacknowledged foundation of political economies based on human labor.

The idea of natural capital exists within a more general theory of natural economy. A natural economy constitutes a system of already existing material goods and services that make possible the production of additional wealth. These goods and services are generally found (in nature) rather than themselves produced by humans. In this sense, natural economy is to be contrasted with political economy, where markets are the mechanisms for distributing capital goods produced by human labor. But natural goods must also be distributed for the production process based on human labor to get going. The market, however, has *not* been historically the main means of distribution in this case. Natural goods such as land have usually been acquired more directly by users, often acting together in local associations of various kinds.

For this reason, as ecological economists emphasize, the natural economy is the necessary basis for (all) human wealth. Without adequate natural wealth (or capital), human (and other) life is not possible (Heal 2016). In this sense, natural economies constitute systems for conserving, rather than depleting, natural goods (as political economies oriented to a constantly increasing stock of economic values might do).

Natural economies do this by their use of cyclical production processes that recycle natural materials so as to replenish the biomass that is the source of these materials. In this way, natural economies provide the means to avoid a "throughput economy where enormous quantities of natural inputs are finally transformed into enormous quantities of . . . waste" (Sachs 1999, 176). The point of regarding land as natural capital, then, is to avoid the view that regards land primarily as physical material to be commodified for exchange in market economies.

One consequence of this latter view is that "land grabs" (to produce profits from land revaluations) can also produce scarcities by depleting the natural economy of material throughputs to be rendered into productive wealth by human labor (Daly 1996, 78–79). These "natural" scarcities—the absence of sufficient natural capital—are not susceptible to redress from egalitarian redistributions of wealth, since they result from limits on the total resources available for human survival.

Furthermore, maximizing efficiency—a standard goal of traditional economic theory and business practice—aims only at the optimal use of *available* resources. A maximally optimal economy is one in which no more improvements in allocations to particular individuals or parties can be made without worsening allocations for others. Aspiring to reach this state of productive efficiency does not, however, imply anything about whether there is a sufficient material basis—that is, natural capital—for such a scale of resource use.

Increasing economic efficiency, and redistributing the proceeds of a more efficient economy, does nothing to prevent greater scarcities of the basic (environmental) goods that will determine what can be produced. In particular, as Herman Daly notes, market efficiencies and allocations—even if optimal—do nothing to limit, and indeed are more likely to maximize, resource use:

> [Pareto-optimal] allocation can be achieved at any scale of resource throughput, including unsustainable scales . . . . Ecological criteria of sustainability . . . are not served by markets. Markets singlemindedly aim to serve allocative efficiency. Optimal *allocation* is one thing; optimal *scale* is something else.
> *(Daly 1996, 32).*

Without the ability to obtain "cheap nature," including inexpensive land, due to scarcities resulting from overexploitation for profit, capital accumulation becomes increasingly restricted (Patel and Moore 2017). For this reason, natural capital, in land and in other physical materials, remains the foundation of wealth in all societies. Preserving natural capital accordingly requires different concepts and principles than those—such as property ownership—that have been used to distribute economic wealth.

One such concept is the idea of *common property regimes*, or, more generally, "the commons." The commons is a term that refers to traditional forms of communal property that predate the modern dichotomy between private and public (or state) ownership. Today, communal property forms seem to some theorists particularly appropriate for natural goods, such as land, that are found, rather than made. In general, these goods are designated as "common-pool resources" (CPRs). Typical examples of CPRs are the oceans and atmosphere, as well as, on smaller geographical scales, lands and waters, to the extent that they function as soils, (waste) sinks, forests, or fisheries (Dolsak and Ostrom 2003, 3–4).

One characteristic of CPRs is that they may be hard to parcel out to users since they often consist of stocks (e.g., fish stocks) with high degrees of variability and uncertainty as to the "flow" of the stocks over time (Dolsak and Ostrom 2003, 9). This problem may occur even with land and water to the extent that useful amounts of such resources only exist at scales large enough that individual owners would have to engage in employing extensive quantities of labor or capital in order to profit from the use of the resources.

There has been considerable debate over the best form of property ownership for the use and management of these resources. Typically, the debate has been over whether private or public ownership of CPRs is the best way to manage them. But as various scholars have pointed out, this is an oversimplification of the types of ownership available. Joan Martinez-Alier, for instance, distinguishes four types of property that can be used to manage resources: [1] private (i.e., individual) property, [2] public (i.e., state) property, [3] community property, and [4] open access (that is, unowned resources) (Martinez-Alier 2002, 74–75). The third type is today usually referred to as common property institutions (CPIs).

It is vitally important to note the differences between CPIs and open access zones (OAZs). Both have been regarded as "commons." One view is that open access zones *are* commons since they lack both private and public ownership as such. In other words, CPIs do not exist. Another view, in contrast, regards CPIs as commons, and open access as denoting their absence (or the decline of CPIs formerly managing these zones). On the first view, commons are unowned, and require either private or public management in order to save the resources available there. On the second view, the absence or loss of such common property institutions means that natural goods are open to all users. If commons are regarded as being unowned and therefore equivalent to OAZs, the reason is the lack of human habitation that would have given rise to either private or public property ownership. If OAZs are regarded as resulting from the *loss* of a commons, this could be the result of private or public interventions to dismantle CPIs.

CPIs are a type of mixed private-public ownership that consists of property jointly held by a group of users. It is difficult to generalize about CPIs, since there are a variety of them, dependent on the historical and geographical features of the resource in question. From village commons lands to international fishing grounds agreements, CPIs generally attempt to do two things: first, they allow *general* access to the resource by all "established" (e.g., traditional or repeated) users—but not *universal* access by anyone. Second, they mandate and enforce conservation measures by the community (of users) to maintain the resource for future use (Ostrom 1990, chap. 3). The role of such property institutions has been historically important in maintaining sustainable use of resources by inhabitants of ecological regions who are dependent on them for their livelihood (Radkau 2014, 16–17).

As Ostrom has shown, CPIs have played a more useful role than either market or state institutions in preserving natural resources over many generations:

> Neither the state nor the market is uniformly successful in enabling individuals to sustain long-term, productive use of natural resource systems. Further, communities of individuals have relied on institutions resembling neither the state nor the market to govern some resource systems with reasonable degrees of success over long periods of time.
>
> *(Ostrom 1990, 1)*

A number of important examples illustrate the ability of CPIs to manage resources sustainably. Fisheries, for one, have historically been maintained in this way: from data gathered in the 1980s, 90 percent of fishing, and over 50 percent of fish consumed occurred and came from "small-scale, inshore fisheries," managed by CPIs (Ostrom 1990, 27). As for forests, research on the most successful example of sustainable forestry management, that of the Japanese islands, indicates that success was the result of community management of local forests (Radkau 2014, 21–22).

What such examples challenge is a received view that local and community control is an unreliable way of managing resources, because of the dangers of free-riding in such cases. Garrett Hardin, William Ophuls, and others have argued that there is a "tragedy of the commons" that results in overuse and depletion of resources not managed either according to strict private-property or state-property regimes. Commons have been regarded as a flawed way to protect and maintain resources that were not the responsibility of any individuals in particular. On one account, only the parceling out of such resources to individuals holds any hope of maintaining them (Hardin 1968). On another account, it is only a strong state that can mandate strict guidelines for, and where appropriate, prohibitions against, use that can save such resources (Ophuls 1977).

The responses to such views are relatively straightforward. In terms of private property advocates such as Hardin, they seem to have confused resources held in common (CPIs) with resources that are unheld (OAZs). While there is a danger of depletion when open access to a resource exists, this does not apply to common or community-held resources, which are bound by rules and conventions regarding use, and available only to a restricted population of users who have agreed to such rules. Furthermore, when resources are unowned, there is often a tendency to institute some kind of CPI by the users themselves (Martinez-Alier 2002, 76).

As for advocates of state ownership, such as Ophuls, they tend to conflate *state* management with *public* management. In a number of important cases, it is state control that has been the problem. For instance, forestry departments in Japan, Germany, India, and elsewhere, that have actually instituted policies resulting in deforestation, rather than conservation. In certain instances, this was the result of states abolishing CPIs and replacing them with open access, where it had not existed before (Ostrom 1990, 23). The absence of state control does not, historically or at present, mean that there is open access—this depends on whether there is a CPI involved in management of the resource.

Nevertheless, it is important to be clear about the conditions under which CPIs will be successful, as well as those under which they are likely to fail. While CPIs, as Ostrom emphasizes, can operate at different scales, the local scale is important because small communities are more likely to be integrated into surrounding ecosystems (Holling, Berkes, and Folke 1998, 356). This means that there is a structure connecting the community with its natural environment—a structure than can be formalized through a common property regime of some kind.

It is the smaller scale of most traditional CPIs, along with other elements of communal solidarity, that strengthen the ability of the community to enforce property rules, and enable CPIs to last. However, it is possible for such property regimes to operate on larger scales, "if users are organized in nested enterprises or multiple layers" (Berkes and Folke 1998, 8). This will be one of the real challenges in dealing with certain resources that are accessible to a much larger user population—devising CPIs that can operate in these more complicated situations.

Finally, despite the necessity of community institutions for sustainable management of resources, it is the case that institutional mechanisms to enforce rules and control access are also essential. Traditional communal lands and waters have often been regulated by spiritual or folkloric beliefs that can command wide assent from the peoples involved. If these beliefs decline or lessen, other incentives or safeguards will be required. When failure of CPIs occurs, it is often from "institutional failure to control access to the resource, and to make and enforce internal decisions for collective use" (Berkes and Folke 1998, 7).

Two aspects of attempting to operationalize this theory of common property institutions or regimes are important. First, it is essential to emphasize the significance of the geographical *scale* at which rules for common property institutions are applied. When the scale of political power or jurisdictional authority (i.e., sovereignty) does not match the scale of ecosystems and natural resource use, there is a danger. In such circumstances, where states have primary authority over smaller resource catchment zones, there is likely to be conflict—specifically, conflicts over the just distribution of ecosystem goods (Geores 2003, 82). This is because larger-scale governance institutions have multiple, and often competing, constituencies that will seek to use natural goods for different purposes.

One famous example of this has been the conflict over the waters of the Narmada River in India. It is famous because it resulted in the creation of the Narmada Bachao Andolan (the "Save the Narmada Movement," along with its spokesperson, Medha Patkar), which led the longest rural political struggle against the government and big landowners in India since independence. While the upriver peoples in Malwa sought to use the traditional water source for their own subsistence, the downriver landowners in Gujarat sought to dam the river, creating power sources for their more energy-intensive uses (e.g., cash crops). Needless to say, such damming would inundate much of the land upriver. The Indian government viewed both constituencies as needing accommodation, rather than the second group as impinging on the water rights of the first group (Radkau 2014, 143–44).

Second, this example highlights the missing element in the theorization of CPIs by social scientists. It is the matter of resource (land) and territorial *rights* that requires further consideration. In comparisons of claims by individuals, states, and communities, who is *entitled* to have priority in accessing, managing, and controlling lands and other natural resources? While this is largely the subject of Chapter 4, a preliminary point that can be made is that common property regimes have an essential connection to considerations of ecological conservation and sustainability, due to local communities' dependence on particular ecosystem goods and services. Corporations with financial capital, and states with military force, sufficient to acquire land on a large scale do not seek this in order to *preserve* the natural goods in these lands, but in order to exploit them for profit or "development." Such land grabs ignore local attachments to these (home)lands, aiming at their capture *from* local communities.

The Need for Territorial Rights

Aside from the scale of land and resources at which peoples seek to satisfy their needs and secure their livelihood, there is also a "normative" scale of values that is essential for establishing legitimate claims to such things. As we examine different types of legitimacy, it is important to understand why "scale shifting" from one to another is sometimes necessary. At the lowest normative scale is the concept of ownership. Common property is a form of ownership that seeks to distribute property rights to communities (rather than individuals) needing access to lands for satisfying basic needs and ensuring livelihoods.

At a somewhat higher—in the sense of more general—scale are territorial rights. These differ from property ownership rights in that they concern territory as a wider context for achieving self-sufficiency in natural goods and services. But there is a second aspect to the contrast between property and territory, as was discussed earlier. This is the matter of territoriality understood as a collective right over and above that of (common) property. Ownership can take either an individualist or a collectivist form—either as private property or as common property. But both involve proprietary relations that, *at some point*, devolve to personal possession and use, even if only temporarily. Commons, after all, were non-privately owned lands that were available to persons who would directly use a portion of these lands. The use was temporary, but it was still that of an individual user. Commons established a right of inclusion (or put negatively, non-exclusion) for a class of (individual) owners—those able to directly use the natural goods involved (e.g., farmers or fishers, using particular lands or waters).

In other words, common property is different from private property; but it still operates within a legal framework of ownership that is not dissimilar from proprietary rights applicable to individuals. Territory, by contrast, in order to be properly maintained and administered, requires the intervention of some political agency. Such political agency is not only the personal agency of the direct users of the good. There is more involved than just the direct relations between owners and owned. Moreover, territorial rights are claimed and protected not, in the first instance, by legal mechanisms (i.e., litigation), but by political means (i.e., legislation).

In this sense, territoriality is a more abstract relation than common property. It also often operates at a considerably wider geographical scale. Nevertheless, there are good reasons why land (and other natural goods) should be bundled together as territories, rather than simply regarded as common properties of direct groups of users. First, there are the ecological advantages of administering territories that unite whole ecosystems or ecological regions, rather than just specific commons organized around usage of certain localities. For instance, the unification of a territory around a river or watershed, or uniting a large, but sparsely populated prairie or desert region, may bring advantages in terms of the maintenance both of the natural goods in those regions, and of the populations that can live sustainably in them. This may be especially applicable to territories populated by pastoral and nomadic peoples, who might range over a large geographical zone in the course of a year or so.

Second, the advantages of establishing territorial governance can also be considerable. Common property regimes that do not control the territories in which they exist are especially vulnerable to others who may seek to utilize the resources unsustainably. The Narmada River case discussed earlier is an example of this: peoples living on and from the river, but not in control of the watershed, were unable to fully prevent (or to prevent at all, without a huge political struggle) the extraction of riverine resources by other, more powerful, forces. For these reasons, the struggle to maintain common property regimes will often necessarily devolve into conflicts over territory. In this context, the ability to make legitimate claims to territorial rights becomes politically important.

Finally, at the highest "normative" scale, there are claims to sovereignty, which is the most political way in which nature, land, and territory is demarcated and defined. Sovereignty claims, as we will see in Chapters 5 and 6, are how property is secured from "external" agencies that might seek to possess or use it for their own purposes. It is at this scale that the principle of ecosovereignty will be most efficacious, in that it will determine who legitimately has *ultimate* authority to dispose of territory.

After a long time in which economic value and social solidarity seemed to have been detached from agrarian relations and land tenure struggles, the importance of land (and land ownership) is again becoming manifest. These forms of social struggle—for millenia prior to the twentieth century the principal venue of conflict between communities, countries, and social classes—have returned because of the increasing scarcity and of land in the twenty-first century. The industrial era now increasingly seems to have been a temporary hiatus, due to unrealistic expectations of the inexhaustibility of natural resources. In the face of mounting scarcities of crucial materials for the further expansion of industrial infrastructure, it is high time to reconsider the use patterns of land and other natural goods.

But, unfortunately, the entrenched wealth and power of a small global elite intent on maintaining their profligate consumption patterns makes it imperative to find new ways in which the political sway of peoples who need access to land for subsistence purposes—and other seeking more sustainable forms of livelihood—can be increased. Most essential is for such peoples to keep or take back control of territories necessary for their livelihoods. The need for land therefore inevitably becomes a political struggle over territory. In such struggles, it is not always clear where the lines are to be drawn. For instance, whether or not the urban poor align with the political elites committed to further industrialization, or whether they reject industrial capitalism as an unsustainable form of livelihood and reorient themselves in alliance with agrarian communities, will be critical to the success or failure of the struggle between these two political blocks.

All that philosophers and theorists can hope to contribute to such conflicts is some clarity about what legitimacy the different parties to these struggles may have. When applied specifically to the matters of land and territory, this translates into the question of who has legitimate rights to which territories, and why. The upshot of this chapter is that those fighting to use land for subsistence and livelihood purposes have a better claim to ownership and control over lands than those merely seeking to extract maximum economic value from them. But whether such claims can generate a definite right to territory is another matter: the history of different sorts of territorial rights must be more thoroughly considered, and an argument for an ecosystem-based territorial right given. That is the task of Chapter 4.

References

Abizadeh, Arash. 2012. "On the Demos and Its Kin: Nationalism, Democracy, and the Boundary Problem." *American Political Science Review* 106: 867–82.

Agnew, John. 1987. *Place and Politics: The Geographical Mediation of State and Society.* Boston, MA: Allen & Unwin.

Agrawal, Arun. 2000. "'Community' and Natural Resource Conservation." In *Nature, Production, Power: Towards an Ecological Political Economy*, edited by Fred P. Gale and R. Michael O'Gonigle. Cheltenham: Edward Elgar.

Berkes, Fikret, and Carl Folke. 1998. "Linking Social and Ecological Systems for Resilience and Sustainability." In *Linking Social and Ecological Systems: Management Practices and Social Mechanisms for Building Resilience*, edited by Fikret Berkes and Carl Folke. Cambridge: Cambridge University Press.

Dahbour, Omar. 2003. *Illusion of the Peoples: A Critique of National Self-Determination.* Lanham, MD: Lexington Books.

Daly, Herman. 1996. *Beyond Growth: The Economics of Sustainable Development.* Boston, MA: Beacon Press.

Deudney, Daniel. 1996. "Ground Identity: Nature, Place, and Space in Nationalism." In *The Return of Culture and Identity to IR Theory*, edited by Y. Lapid and F. Kratochwil. Boulder, CO: Lynne Rienner.

Dolsak, Nives, and Elinor Ostrom. 2003. "The Challenge of the Commons." In *The Commons in the New Millenium: Challenges and Adaptations*, edited by Nives Dolsak and Elinor Ostrom. Cambridge, MA: Massachusetts Institute of Technology Press.

Dovers, Stephen, and Colin Butler. 2015. "Population and Environment: A Global Challenge." *Australian Academy of Science*. science.org.au (online).

Fairbairn, Madeleine. 2020. *Fields of Gold: Financing the Global Land Rush.* Ithaca, NY: Cornell University Press.

Folke, Carl, Fikret Berkes, and Johan Colding. 1998. "Ecological Practices and Social Mechanisms for Building Resilience and Sustainability." In *Linking Social and Ecological Systems: Management Practices and Social Mechanisms for Building Resilience*, edited by Fikret Berkes and Carl Folke. Cambridge: Cambridge University Press.

Geores, Martha E. 2003. "The Relationship Between Resource Definition and Scale: Considering the Forest." In *The Commons in the New Millenium: Challenges and Adaptations*, edited by Nives Dolsak and Elinor Ostrom. Cambridge, MA: Massachusetts Institute of Technology Press.

Hardin, Garrett. 1968. "The Tragedy of the Commons: The Population Problem Has No Technical Solution; It Requires a Fundamental Extension in Morality." *Science* 162 (3859): 1243–48.

Heal, Geoffrey. 2016. *Endangered Economies: How the Neglect of Nature Threatens Our Prosperity.* New York: Columbia University Press.

Holling, C. S., Fikret Berkes, and Carl Folke. 1998. "Science, Sustainability, and Resource Management." In *Linking Social and Ecological Systems: Management Practices and Social Mechanisms for Building Resilience*, edited by Fikret Berkes and Carl Folke. Cambridge: Cambridge University Press.

Honoré, A. M. 1961. "Ownership." In *Oxford Essays on Jurisprudence*, edited by A. G. Guest. Oxford: Oxford University Press.

Hornborg, Alf. 2012. "Accumulation: Land as a Medium of Domination." In *Ecology and Power: Struggles Over Land and Material Resources in the Past, Present, and Future*, edited by Alf Hornborg, Brett Clark, and Kenneth Hermele. London: Routledge.

Kingsnorth, Paul. 2017. *Confessions of a Recovering Environmentalist, and Other Essays.* Minneapolis: Graywolf Press.

Kolers, Avery. 2012. "Justice, Territory and Natural Resources." *Political Studies* 60: 269–86.
Martinez-Alier, Joan. 2002. *Environmentalism of the Poor: A Study in Ecological Conflicts and Valuation.* Cheltenham: Edward Elgar.
Moore, Margaret. 2015. *A Political Theory of Territory.* Oxford: Oxford University Press.
Nine, Cara. 2008. "A Lockean Theory of Territory." *Political Studies* 56: 148–65.
Nine, Cara. 2012. *Global Justice and Territory*. Oxford: Oxford University Press.
Nine, Cara. 2022. *Sharing Territories: Overlapping Self-Determination and Resource Rights.* Oxford: Oxford University Press.
Nozick, Robert. 1974. *Anarchy, State, and Utopia*. New York: Basic Books.
Ophuls, William. 1977. *Ecology and the Politics of Scarcity: Prologue to a Political Theory of the Steady State.* San Francisco, CA: W. H. Freeman.
Ostrom, Elinor. 1990. *Governing the Commons: The Evolution of Institutions for Collective Action.* Cambridge: Cambridge University Press.
Patel, Raj, and Jason W. Moore. 2017. *A History of the World in Seven Cheap Things: A Guide to Capitalism, Nature, and the Future of the Planet.* Berkeley, CA: University of California Press.
Pimentel, David, and Michael Burgess. 2013. "Soil Erosion Threatens Food Production." *Agriculture* 3: 443–63.
Purdy, Jedediah. 2019. *This Land Is Our Land: The Struggle for a New Commonwealth.* Princeton, NJ: Princeton University Press.
Radkau, Joachim. 2008. *Nature and Power: A Global History of the Environment.* Cambridge: Cambridge University Press.
Radkau, Joachim. 2014. *The Age of Ecology: A Global History.* Cambridge: Polity Press.
Relph, Edward. 1996. "Place." In *Companion Encyclopedia of Geography*, edited by I. Douglas, M. Robinson, and J. Williams. New York: Routledge.
Sachs, Wolfgang. 1999. *Planet Dialectics: Explorations in Environment and Development.* London: Zed Books.
Scruton, Roger. 2012. *Green Philosophy: How to Think Seriously About the Planet.* London: Atlantic Books.
Simmons, A. John. 2001. "On the Territorial Rights of States." *Philosophical Issues* 11: 300–326.
Steiner, Hillel. 1996. "Territorial Justice." In *National Rights, International Obligations*, edited by Simon Caney, David George, and Peter Jones. Boulder, CO: Westview Press.
Ypi, Lea. 2012. "A Permissive Theory of Territorial Rights." *European Journal of Philosophy* 22: 288–312.

4

AN ECOSYSTEM-BASED RIGHT TO TERRITORY

Land, Ecology, and Territoriality

Territorial rights are high-priority claims of collective human agents (peoples) to material (natural) goods (and services) under certain conditions. What needs further specification is: [1] the proper understanding of the material/natural goods in question (territory), [2] the nature of the agents (peoples) entitled to claim control over territory, and [3] the relevant conditions under which such claims are legitimate. An outline of the argument given here is as follows. First, territory involves not only the human valuation of land but also a valuation based on the proper understanding of land as a natural good. This means that territory is defined as land that is ecologically essential for human (as well as nonhuman) livelihoods. Second, there is a "conceptual connection" between territory understood in this way and the assertion of peoplehood by social groups with a particular relation to land. This connection involves the co-definition of particular territories and peoples, in relation to one another. Third, legitimate claims to territory (and control, or rights, over it) depend on establishing and maintaining a stewardship relation to the natural goods inherent in that territory.

Territoriality is the feature of land that makes it a territory. This feature is twofold: on the one hand, it is the result of the use of land for human habitation; on the other hand, it is the value of land as a means toward human livelihood. So territory is both closely connected to peoples and their identification as distinct social groups, and to ecosystems as they provide materials for human social reproduction. Avery Kolers, in his book, *Land, Conflict, and Justice*, offers some useful specifications of territory related to these considerations. Kolers maintains that the "global climate emergency" mandates putting "environmental

DOI: 10.4324/9781003568155-7

sustainability—particularly stewardship of the climate and of ecosystem services that sustain civilization as we know it—at the core of legitimate state territorial claims" (Kolers 2009, 3). Kolers' work has been one of the first to clearly highlight the relation between the human/political and the natural/environmental aspects of territoriality. This is surely a great advance in the treatment of territory as a politically important category. This contribution is, however, somewhat diminished by his idiosyncratic approach to the use of geographical concepts in his articulation of what territory means.

Territory needs to have some relation to the concept of peoples, since this term is itself a way of indicating the difference between territory and other related terms, such as land or property. As Kolers writes, this means that we need to articulate a "political theory of territory" in order to do justice to this aspect of territoriality. Yet, he is reluctant to use the concept of people as such; instead, he substitutes the term, "ethnogeographic community." Here is the role the term plays in Kolers' conception of territorial rights: "A territorial right exists if and only if an ethnogeographic community demonstrably achieves plenitude in a juridical territory; this right grounds independent statehood only if there is no competing right and the territory is a country" (Kolers 2009, 5).

The term, "plenitude" indicates that Kolers emphasizes the environmental, more than the political, aspect of territoriality. Plenitude is defined as the abundance of natural goods that sustain life in a place, including a high degree of diversity in land use. It is the underpinning for a "resilient" society, resilience being "an ecological concept denoting the capacity of a system to bounce back to an equilibrium" (Kolers 2009, 4). These concepts prepare us for the main claim that Kolers makes about territorial rights: "Territory is held in trust for a population, including future generations and future immigrants; thus territorial rights have at their core a *stewardship element* that is typically absent from or peripheral to property rights" (Kolers 2009, 27).

The idea of environmental stewardship is an important aspect of territoriality because, in order to establish an attachment to territory that is more than an elective one, there must be something intrinsic to the places that peoples claim as their own. The ecological/environmental aspect of territoriality, in other words, needs emphasis due to the tendency in much contemporary political theory to assume an abstract concept of space, rather than a notion of place indicating essential connections of some kind between peoples and the countries they inhabit. For many political theorists, in other words, there is nothing *special* about the particular geographical locations of peoples, countries, or states—they are just where they happen to be. Some philosophers even attribute an aspect of (moral) arbitrariness to the particular places where peoples live—they might as well have been elsewhere (Beitz 1999). The thought, put this way, is bizarre: in another place, there might be another people, another country, another state—but never the *same* one. This rather obvious point is missed or forgotten in much political theorizing about sovereignty and territoriality. But how are we to characterize the nature of the specific connection of peoples to places? Only by establishing a much closer identity between them: peoples are *who* they are, in part, because of *where* they are.

Kolers' way of making this point, however, does not quite yield what I refer to as the conceptual connection between peoples and territories. Rather, he argues that peoples or groups, variously defined, always possess an at least implicit "ontology of land." These are "conception[s] of land" shared by a community that results in their practicing particular land-use patterns. For instance, "Anglo-American ethnogeography" has an ontology of land that involves single-family properties, home-work separation, and an automotive-centered transportation infrastructure. Other ontologies that Kolers mentions include agrarianism (family farms) and pastoralism (livestock grazing) (Kolers 2009, 86).

This way of characterizing land use, and by extension, territoriality, is useful, up to a point. Such "ontologies" are described by Kolers as based on "concrete relationships," "material rather than intentional," and "unencumbered by the psychological baggage of identity" (Kolers 2009, 91–92). But there are two problems with this approach. On the one hand, it is a relativizing one that undercuts the judgments that Kolers seems to want to make about the difference between sustainable or resilient patterns of land use and other ones. On the other hand, such a view of land use does not seem able to do justice to the conflictual—that is, political—nature of territoriality; it does not make clear the basis for legitimate or illegitimate claims to territory.

The first problem results from the use of "ethnogeography" as a way to define peoples, and their connections to territories. The term is derived from work by J. M. Blaut on the geography of national groups (Blaut 1979). According to Blaut, *every* human group has an ethnogeography, involving a "belief-system" (about land use) (Blaut 1979, 6). Yet, if that is the case, how are they to be distinguished from one another? In particular, how is it possible, as Kolers seems to want to do, to reject certain territorial claims due to the unsustainability of the type of land use to which such claims would lead?

For instance, he argues that "Anglo-American ethnogeography" suffers from an "inability to recognize the value of places, ecosystem services, and land for anything but economic exploitation . . . . [It] treats land as wholly passive and merely a commodity" (Kolers 2009, 134–35). But Kolers' theory implies that, if there is an ethnogeographic community, then they have an "ontology of land" that, however objectionable from an environmental point of view, nevertheless is a viable relation to a territory. In other words, how is it possible to *criticize* a view of territory, if it is rooted in a "belief-system"?

The environmentally oriented aspects of Kolers' theory of territory therefore seem at odds with the "ethnogeographic" (or perhaps, anthropological) basis of the theory. The first aspect is the more constructive part, and contributes to an understanding of the ecological aspects of territoriality. For example, territories or countries are sustainably maintained when they are or become resilient (Kolers 2009, 76). Resilience here means redundancy; too efficient use of resources may lead to inability to recover from disturbances to ecological equilibrium in the face of unanticipated setbacks and disturbances (Kolers 2009, 113–15). But the second part of this theory, ethnogeography, does not substantiate such conclusions. They accord better with more recent work in geography on scale and its relation to distinctive forms of land use and territoriality (Sheppard and McMaster 2004).

One of the interesting consequences of regarding territoriality at least in part as a function of resilience is that it can suggest the limits of sustainable territories. Territories may be too small to be resilient, in the sense that they are not diverse enough to allow for redundancies and multiple uses. They may also be too large, in the sense that they become too dominated by a single pattern of land use, as a result of the search for comparative advantage by one or another subterritory. Nevertheless, despite the advantages of using the ideas of plenitude, resilience, and sustainability in defining territory, it is questionable whether this approach can provide criteria for resolving territorial *disputes*.

This suggests that the other problem with Kolers' ecological concept of territory is that it does not actually generate territorial *rights*, in the sense of legitimate claims by peoples or groups to control over particular territories. This is the case despite Kolers' claim that his approach provides a way to "include a close link between ecological and political principles" (Kolers 2009, 136).

Margaret Moore, whose work we will examine later in this chapter, makes this point by arguing that the principle of plenitude, which is supposed to provide a way of distinguishing between rival claims to territory, is too vague and indeterminate to do this work (Moore 2015, 121–22). The problem with Kolers' theory of territoriality is that it regards the connections of peoples to territories as dependent upon ethnological identities which generate affinities to particular environments. It is assumed that there are many "ontologies of land," several of which are "plenitudinous." Which, if any, can claim a right to realize their conception of land use?

The problem here goes back to the "ethnogeographical" concept underlying this view. Relying on a narrow and unscientific concept of human geography—that the connection of peoples to lands is based on ethnonational identities—it is unable to generate definitive conceptions of legitimate and illegitimate territorial claims. This is despite the useful emphasis on resilience as a criterion of sustainable land use. The foundation of territorial claims certainly turns out to be conflictual—since different national groups have often clashed over rival territorial claims—but not in a way that can provide ideas for resolving such conflicts. For this, we need to look more closely at the political dimension of territoriality. But first, some consideration should be given to why ethnonational identities cannot ground legitimate territorial claims. This is the case despite a common understanding (including by Kolers) of people-land connections as mediated by the identities of national groups.

Against Nationalism

Linking territoriality with nationality is difficult in a couple of respects. First, national identity has been used to justify certain political entitlements such as self-determination or secession, but not usually a right to territory as such (Dahbour 1997). Second, territory finds its justification more readily in individualist or collective property rights to land, as was discussed in Chapter 3. National identity seems to concern, rather, the political amalgamation of persons of the same ethnicity into nation-states, rather than into territorial entities variously defined. This is because the notion of national identity that underlies claims to nation-states relies on the identification with others in a collective—a national group. This may, or may not, involve the attachment to a place that seems essential for making a territorial claim. Daniel Deudney has put this difference evocatively by distinguishing the "we-feeling" important to nationalism that involves "group solidarity and attributes shared in common" from a "here-feeling" that is "derived from shared habitation of a place" (Deudney 1996, 130).

Equating nations with peoples seems to be the obvious way to make the connection between nationality and territoriality. But this connection is not easy to make since the idea that peoples just are nations and nothing else has its own distinct difficulties (Dahbour 2003, 18–42). Nevertheless, this idea has been put forward by Michael Walzer, David Miller, Bernard Yack, and other exponents of national rights, most of whom claim the term, "nation," denotes all or any non- or pre-political groups or collectives—to the extent that they also have political aspirations. Here is how Yack puts it: "The nation provides precisely what is lacking in the concept of the people: a sense of where to look for the prepolitical basis of political community" (Yack 2001, 524).

The problem with this view is that regarding peoples simply as nations does not provide the linkage that the latter concept requires between territoriality and statehood—that is, the spatial location of peoplehood. National groups are inevitably linked to states, because their political claims just are to (become or possess) a (nation-)state. But what is the connection to territory? As Margaret Canovan notes, "The geographical metaphor of 'the country' [is] a term closely linked to the nation" (Canovan 2005, 96). But it is not the same. Countries are territories inhabited by specific peoples; these peoples need not be national groups. In fact, they rarely are, because settlement patterns and population movements inevitably complicate attempts to achieve enough national homogeneity to claim a basis for a nation-state. Even in the canonical European states created in the era of nationalism—Italy and Germany (not to speak of France or Britain in an earlier age)—there was no national homogeneity among the inhabitants of these countries at the time of political unification. Whatever these European states were, they were not nation-states. Nationalism is a political project—the project being to transform a country (and its people) *into* a nation (Breuilly 1982). Needless to say, this implies that the two are not the same.

Nevertheless, it is possible to connect the theory of national rights with the matter of territory. This connection, though, will involve a number of philosophically dubious moves designed to justify certain equally questionable political claims. In the end, a nationalist conception of territory does not yield a theory that can justify territorial rights so much as an ex post facto validation of the claims of historically and politically dominant groups. The example of Zionist claims to Palestine (and, in fact, to "Greater Israel") is an important case of this. The moves taken to justify such claims are the following: [1] a valuation of territory in terms of its role in the creation of national identities; [2] the justification of historical claims to territory based on these identities; and [3] a right of certain national groups to specific territories justified by their political control of them. These moves can be made in more or less strident ways; both versions will be considered.

First, what is the value of territory for nationalists? As Tamar Meisels notes, territory is viewed by nationalist thinkers as a "constitutive component of [a nation's] identity" (Meisels 2005, 107). On her view, the value is accrued through the settlement, and subsequent "development," of territories. Second, the nationalist valuation of territory is historical, in the sense that it is the result of historically accrued connections or attachments between a national group and a specific territory. The national group can exist separately from the territory—and in fact, these are the cases in which a territorial right becomes important. It seems, then, to be the history, more than the constituted character, of a nation that generates claims to territory. If nations are constituted by elementary features of social life, it would seem to be the attachments to an identity marked by ethnic singularity, or *consanguinity*. Yet, the feature that clearly characterizes peoples who are intrinsically connected to territories is that of *contiguity* (Dahbour 2003, 31–34). National groups, in other words, are better characterized anthropologically—by their real or imagined social relations—not geographically, as a result of their spatial locations. This is why national groups that have been expelled from territories can (sometimes) retain their identities: the connection to territory is less important than the kinship and other connections that maintain the identity of the group (Horowitz 1985).

Furthermore, to the extent that the nationalist view of territory is constituted by ethnicity and history, this is often the result of *beliefs* about the matter, not facts. To the extent that connections between national groups and territories are regarded as established, they are established not by actual historical relationships, but by the belief that there were such relationships. Examples abound;

one is Israel, to be discussed shortly. Another is Greece, where the connections between modern Greeks and the territory of the modern Greek state and contiguous regions (claimed by Greek nationalists) is tenuous at best (as is their alleged connection to the ancient Greeks) (Agnew 2009, 69–78). The inherent subjectivism of the nationalist connection to territory vitiates the idea that this connection has a territorial basis; it is founded, rather, on beliefs about certain places that may have little relation to either the actual history, or the lived experience, of existing peoples in those places.

These problems with the basis of the nationalist view of territory has consequences for rights claims based upon the view. Meisels postulates two kinds of territorial rights, both "historical," in her terms: [1] a "First Occupancy Right," and [2] a "Right to Formative Territories" (Meisels 2005, 30). The first right is similar to a Lockean claim of property ownership and seems extraneous to the main argument. This main argument involves the idea of the second right, one that is based on a right to those territories that are important for the "formation" of national identities.

But what if these two rights were to conflict, for instance between one group that occupies territory which is not constitutive of their identity, and another group that does not occupy it, but nevertheless claims it as constitutive? The idea that territory is valuable mainly for its role in constituting (national) identity suggests that the second right would override the first. Certainly, this would have to be the case if the theory were to retain ties between national groups and territories. So the right is one of groups claiming territory important for their identities. Now, what if two such groups claim the same territory, for the same reason (e.g., Palestine/Israel)? How would this view of territorial rights resolve conflicts based on such contending claims to constitutive/historical connections to land?

Any resolution would result from legitimating the dominance of national groups that occupy, settle, and develop territories in accordance with their beliefs about the importance of the territory for their national identity. Here is how Meisels puts it: "Where members of a nation have *already succeeded* in settling a parcel of land and in shaping it [in a way that] reflects their culture, there are strong moral reasons to grant their nation a right to that land" (Meisels 2005, 95; italics added). "Success" is crucial here: the nation that gets and keeps the land has the right to it. The nation that is unable to hold onto territory loses the right to it. But there is a further quandary: How could a nation that does not already occupy territory ever have a legitimate claim to it?

Meisels raises this problem herself in the course of discussing Zionist claims to Palestine. She argues that the Zionist case is paradigmatic for assertions of the right to *lost* territories (Meisels 2005, 58). But how to justify the reclamation of such lost territories, if occupancy by another national group (e.g., Palestinian Arabs) is ongoing? The first step is to establish an "historical" connection between a national group and a territory. The way to do this is to assert its importance for a particular understanding of the identity of the group. Zionism, she maintains, is a case "in which people were separated from their historic territories (or parts thereof) by parties other than the present occupants" (Meisels 2005, 43). This suggests that, were the people who had lost the territory able to succeed in reoccupying the land, they would have the right to retain it. This goes for territory not yet reoccupied as well, of course. This is where "Greater Israel" (i.e., various territories beyond Palestine) comes in: the ability to (re)occupy these territories would also generate a right to them, on this view.

Is there any way to moderate this nationalist conception, so as to avoid entailing the legitimation of assertions of force by national groups? David Miller attempts to do this by focusing in his account of a nationalist concept of territoriality on the "transformation" of territory, a feature that was mentioned but not emphasized by Meisels. In order for a national group to claim a right to territory, it must have "added value" to the land in question. The value is of two kinds—material (i.e., economic development) and symbolic (i.e., "rituals and practices"). A national group must find and/or produce value of both kinds in order to have a valid claim to territory (Miller 2012, 258–59). There are two further conditions. First, another national group must not be displaced by the assertion of the right; and second, the symbolic value attached to the territory must be of a degree to regard it as a "homeland," not simply as a place where one or two events occurred of "national" significance (Miller 2012, 261).

Presumably, these conditions are introduced to avoid justifying cases of "settler colonialism" (such as Israel). Nevertheless, the problem with this argument is that it is not clear why a national group would not have a right to territory if they only attached symbolic meaning to it, whether or not anything else was done. Miller struggles with this problem, arguing that a group that reduced the material/economic value of land, even though they attached symbolic significance to it, would forfeit their right (Miller 2012, 260). But it is not clear on what basis he makes this argument, since no principles are introduced to constrain or otherwise judge specific uses of land and resources.

In general, "liberal nationalist" arguments such as Miller's suffer from an ambiguity in attempting to combine nationalist and non-nationalist conceptions of identity. This, as Margaret Moore points out, corrupts Miller's account of what constitutes a national group; and it also undermines his argument that national groups must satisfy an additional criterion before they can lay claim to territory (Moore 2015, 79–80). On what basis could Miller argue for denying a national group the right to their territorial homeland simply because he does not agree with the way they seek to develop its resources? If the transformation of a territory—for purposes of "development"—is the key thing that justifies a territorial right, then why limit it to national groups, as opposed to other peoples who also have development as a goal? Conversely, if territorial rights are fundamentally national rights, why deny a national group a territory because they have not (yet) transformed it in a way that makes their claim to it seem permanent?

Miller's account of territory is more congenial than Meisels' in avoiding the implication of "might makes right." But it is also a more incoherent account, for that reason. Justifying territorial rights (or other political claims, for that matter) on nationalist grounds requires prioritizing national groups' self-understanding and beliefs about location and territory, whether or not they, at any given time, have succeeded in inscribing such beliefs into the land and landscape of that territory. The right, in other words, is justified by the beliefs, not the actions, of national groups—since national identity is fundamentally a function of beliefs (well- or ill-founded), anyway. Since such beliefs engender conflicts with other national groups, or with other peoples not constituted by national identities, a national claim-right to the territory produces, rather than resolves, conflicts (Dahbour 1998, 2003, ch. 6). In addition, it does nothing to prevent, and may contribute to, extractivism in the service of nation-state building.

The Conceptual Connection Between Territoriality and Peoplehood

Regardless of these problems, the nationalist attempt to justify territorial rights does point to a valid question concerning the exact nature of the linkage between peoples and territories. If national identities that valorize such linkage on the basis of beliefs about lands are founded on illusory claims about the intrinsic or historically significant nature of such linkages, what would be a better way to understand the connection that links peoples and territories?

The argument of this chapter is that there is a *conceptual connection* between territoriality and peoplehood. That is to say that, in order to define territoriality, a conception of peoplehood is needed, and in order to define peoplehood, territoriality must be referenced. The two concepts mutually imply one another. This internal, or conceptual, connection between the notions of territories and peoples, I refer to as the *conceptual argument*, by which territories and peoples are co-defined (Pateman 1985, ch. 6).

To understand this conceptual argument, we need to see how territoriality, though derived from concepts of place and land, as detailed in Chapter 3, is nevertheless a spatial concept that links physical spaces to the existence of peoples in those spaces or places. Territory has a relation to land that, in denoting spaces with names and boundaries, indicates that they are, in fact, places. Territory is further to be distinguished from territoriality.

Territory is a "bounded space" that has a relation to such notions as "terrain" or land (Elden 2010; Sassen 2013). It is an "intentional kind" that results from intentions to delimit a certain physical space, for various purposes (Kolers 2012, 277). The question is: What purposes? This is where territoriality comes in. Territoriality can be understood as the desire or search *for* territory. As Robert Sack puts it, territoriality is "a *strategy* for influence or control . . . over a geographic area" (Sack 1983, 55). This is a broad definition; many different intentions or purposes would qualify. The narrower definition is given by Saskia Sassen, when she writes that territoriality is a "legal construct" marking state authority over a "territorial jurisdiction" (Sassen 2013, 85–86).

So which is it: a general striving for territory, or a specifically state-oriented or legal demarcation of (a) space? James McCarthy provides the crucial missing element in both these definitions—the search for power or authority (over territory) by political actors, state or non-state: "Territoriality is the development and exercise of power through control or influence over a bounded space and its contents, including both population and resources" (McCarthy 2007, 958). The reason this is a useful definition is that it does not pre-judge who the political actors are who are seeking this power. As McCarthy writes, "It is vital to recognize that territoriality is also a feature of nonstate actors." Definitions such as Sassen's that assume that territoriality is a function of *state* power assume that "the state is the natural or only scale of territoriality" (McCarthy 2007, 959). Conversely, general definitions such as Sack's assume that any purpose, including cultural or economic ones, can motivate efforts to control territory. But it is important to recognize that it is purposes specifically related to the collective intentions of peoples that constitute territoriality.

Territoriality, then, constitutes the intentions of groups, commons, movements, states, *and above all peoples,* to seek power over bounded spaces that they can control for their own purposes. In the case of peoples, these intentions are to seek authority over territory necessary to enable them to live collectively—that is, to embody peoplehood in specific places. Territoriality is a name for the intentionality of peoples who seek to define themselves in relation to some territory, so as to be able to have a collective life *in those particular places*.

Viewed from the perspective of peoplehood, territory is an enabling condition of becoming and remaining a distinctive people. But what is that? Peoples cannot simply be defined as aggregates of individuals, since this will fail to capture many connotations of the term—above all, its territorial component. Similarly, regarding the people as the body of subjects of a ruler is anachronistic in the sense that sovereignty has become disassociated from monarchical or authoritarian rule. Furthermore, peoples retain meanings distinct from such terms as "citizenry" or "nation" (Canovan 2005, 2–4).

The distinction of peoples from bodies of citizens is a relatively easy one to make. There are peoples (e.g., the "German people") who predate the establishment of states; and there are peoples (e.g., the "Polish people") that survive "state death" (Moore 2015, 28). But there is a difficulty here, inasmuch as the concept of peoplehood is conceded to be, in some sense, a political one. Peoples seem to be political entities that are not identical to the citizenries of states—but what sort of entities? One response is to identify the political character of peoples with the very ambiguity of its meaning. In other words, what makes peoples political just is its contested and contestable nature. This is Margaret Canovan's view: the people is an essentially political notion because its "characteristic features mirror the contingency of politics itself" (Canovan 2005, 140).

But this view seems to allow some skepticism about whether the term has a discernible meaning at all. Canovan rightly rejects this skeptical view because it makes the questionable assumption of a reductive individualism that would deny that (social) groups of any kind (e.g., citizenries, national groups, social classes), not only peoples, are real (Canovan 2005, 132). Another possibility, rejected earlier in the text, is that peoples just are national groups. This view also carries the "contingency" of peoples too far in another respect: that they are self-defining in terms of imputed ethnic identities—identities often based on nothing more than ex post facto mythologies about "original" nations.

The indeterminacy of peoples as a concept evaporates if it is linked to territory, as was suggested earlier. Peoples just are groups seeking control over territory, in order to live in those places. But there are two aspects of this definition, only the first of which has been, until quite recently, acknowledged and theorized. The first aspect is the political one of expressing collective intentions toward a territory. The second, and more neglected, aspect is that of the *kind* of intention expressed—namely, that of sustainably inhabiting a territory within its boundaries, and through use (but also, maintenance) of its resources. This is the ecological aspect of peoplehood (and of territoriality).

The first aspect requires specification of the political forms that are best suited to expression of collective intentions *about land*. Without such specification, as Margaret Moore has noted, there is no principled way to settle disputes over territories by states, other than by force; nor is there a way to resolve sovereignty disputes within states between communities or regions (Moore 2015, 2–3, 7). The second aspect requires understanding the best territorial scales for sustainable management of ecological systems. This requirement is made necessary—despite a history of thought about territoriality than ignores this dimension—by the increasing incidence of local and regional environmental catastrophes (e.g., floods and droughts), in addition to the increasing numbers of "ecological refugees" who will exacerbate territorial conflicts in non-catastrophe prone regions (Moore 2015, 241; Nine 2012, ch. 8).

The conceptual argument concerning the mutual implication of territoriality and peoplehood therefore asserts the following propositions. First, territoriality is the struggle for control over territory by a people (not only a state, nor a nation as such). Second, peoplehood is defined by this struggle: peoples are collectives who seek to exert control over territory (not just over individuals, or over other groups). Third, the political aspect of this linkage is that territory is a setting for political action, especially concerning land use and tenure, and that peoples are political collectives focused on particular usages of land and other natural goods. Fourth, the ecological aspect is that territorial rights are justified, and popular rule over territory legitimate, only when the natural goods of particular territories are protected and sustained over time.

The most interesting recent justifications for territorial rights concern these last two points. On the one hand, there are some who maintain that territory is an essentially political concept, and that rights to it can only be justified in terms of its political meaning. On the other hand, there have been some attempts to justify territorial rights, not as an essentially political entitlement, but as based on metaphorical and symbolic understandings of spatial phenomena (such as borders or rivers). Neither of these approaches endorses the conceptual argument that territories and peoples are mutually implicating, nor that the political and ecological elements are both essential to a full justification of territorial rights. The next sections will examine such views, and justify the conceptual argument by showing the linkages of territoriality and peoplehood.

The Political Theory of Territory

To understand the political dimension of territoriality, it is important to be clear about the *purposes* of claiming rights to territory. While, as we will see, the ecological importance of territory mandates a sustainable approach to land use and management, the *political* purpose of territorial rights is distinct. It is the connection to peoplehood—the fact that territory is necessary in order to be a people, in the first place—that is paramount here. So the ways in which territory makes possible the political expression of peoples is what is most important.

Yet, such a conception of territory must avoid some pitfalls in order to understand the true functions of territory in this respect. Above all, it must not prioritize occupancy as such, as nationalist conceptions of territorial rights do. Occupancy, especially when achieved by force, settles nothing about the rights of those claiming the territory. To be clear about that requires understanding the motives, goals, and purposes of claiming territory. Ruling over a territory is to have the power to do—what? That is the question.

In addition, it is important to understand that the political aspect of territoriality is *not* equivalent to the notion of state authority. Much discussion of territoriality—including recent treatments of it—have assumed this equivalence. For instance, Anna Stilz's 2019 book, *Territorial Sovereignty*, which establishes a high standard for the normative defense of the sovereignty concept in general, nevertheless seems to simply assume that sovereignty over territory just is held by states (Stilz 2019). There is no doubt that this is sometimes the case, and that it should sometimes be so, at least in certain respects. But as James McCarthy has cautioned, it is vital to remember that territoriality, even if it is considered to be, in some sense, inherently political, is not simply another name for the control of lands and countries by states.

It is the people, not the state, that, as a general matter, should be regarded as the proper claimant of a right to territory. As Margaret Moore puts it, the dominant view of territory in political philosophy is that it is the "domain of jurisdictional authority," not the possession of a state or ruler. If the people hold that authority (i.e., sovereignty), then it is similarly the people that hold rights over that domain (Moore 2015, 26). But if territory is understood to be the domain of political authority, and political authority requires states, is it not the state that still holds the right to the territory? Moore denies this is the case: "One could conceive of the state as a necessary mechanism by which a group of people could exercise their rights, while the people themselves are the fundamental territorial rights-holding agent" (Moore 2015, 97). The important point is that the basis for the territorial right is the "popular sovereignty ideal of democratic theory" (Moore 2015, 27). This ideal is summarized by Moore as the "collective self-determination of the people."

But before we examine this basis in somewhat more detail, there is an initial problem with this view that Moore must solve. If a territorial right is generated from the collective self-determination right of peoples, then its basis would seem to be a democratic one: the democratic will-formation of peoples includes justifiable claims to the territory upon which this will enacts and administers laws. But the problem is that there does not seem to be any way for territorial limits, boundaries, or borders to be determined democratically. This problem, which has been recognized for some time as a limitation of democratic theory (and which Moore acknowledges), would seem to disable a democratic theory of territory before it can get going.

The basic problem has been outlined by Frederick Whelan: though democracy provides a proceduralist method for resolving political conflicts within a people, it cannot provide such a method for determining *who* is a people (and where they reside, which is the issue with territory) (Whelan 1983, 14–16). Any procedure must occur within definite limits (i.e., involving all and only the members of a given people), and these limits must be antecedently determined, before the procedure can take place. Yet, these limits cannot themselves be determined democratically (since there is no agreed upon population or territory to participate in them). So democratic procedures can only occur within demographic and geographic limits—limits that must be established non-democratically before the procedure can begin.

A standard response to this is to advocate a plebiscitary procedure to resolve conflicts over disputed territories or boundaries democratically. But this is actually only an example of the intractability of the general problem: any plebiscite (a democratic vote to determine sovereignty over a territory) can only occur within boundaries, which themselves may be disputed. To avoid an infinite regress, the boundaries of a plebiscite must be determined non-democratically. As Whelan notes, plebiscites work only when the boundaries within which the plebiscite is to take place are not themselves in dispute (Whelan 1983, 33; see also Dahbour 2013, ch. 3).

Moore takes a different approach: she would not characterize her view of territorial rights as a democratic one, if by that is meant that the right is to be accorded via a democratic procedure. Rather, she argues that the right to territory is a matter of justice—and that the principle of justice in this case is that of political self-determination (Moore 2015, 116). Another way of putting this, however, connects it to democratic theory. That is to see self-determination as naming the substantive good of democracy, and the pursuit of self-determination as the realization of an important value, or the acquisition of a high-priority good. Nevertheless, the question remains: Who is to be self-determining, and where? To answer this question, we must look at Moore's basic view of territorial rights, later returning to see whether her theory can answer the general problem that all democratic theories face: To whom and to where do they apply in particular instances?

Here is Moore's summation of her theory:

> The overwhelming argument for having a right to territory, a right to jurisdictional control over a particular piece of land, follows from the value of collective self-determination, and the locus of that self-determination is determined by having a relationship *both* with other persons and with the land itself, on land that is occupied not unjustly.
>
> *(Moore 2015, 86; italics added)*

There is much to agree with in this concept of territoriality. But there are two questions about the conception that give pause. First, what does a *political* theory of territory mean such that it is different from a nationalist conception (which is, after all, "political," in some sense)? Second, how is the existence of peoples (which is what she means by a "relationship . . . with other persons") and the attachment to territories connected, exactly? The first question is important in order to avoid the quandary of democratic theories in defining peoples (and their territories)—that is, that they *cannot* (initially) be defined democratically. The second question is important inasmuch as peoples may rightly inhabit some territories—but which ones? This is what she means by occupying a territory justly. Furthermore, peoples may have *different* kinds of relationships to territory; is *any* kind of attachment sufficient to generate a right to that territory?

The key to the political theory of territory that Moore is espousing is the idea that a people, *inasmuch as they are a people*, possess a right to the territory to which they are attached. But, if peoples have a right to territory just by virtue of the fact that they are a people, then there must be something about the "collective self-determination" of peoples that makes territory an intrinsic aspect of the act of "self-determining" itself. In other words, the attachment to territory must be built into the definition of peoples such that there cannot be peoples without territories to which they are attached. If this is the case, then to the extent that self-determination is justified by democratic theory in general, the right to territory will also be justified by extension from the very concept of self-determination.

Yet, when we turn to Moore's definition of a people, the attachment to territory has fallen away. She lists three conditions necessary for a people to exist: [1] subjective identification with a "common political project"; [2] the capacity (as in ability to create political institutions) to exercise self-determination; and [3] a history of political cooperation (that would presumably give evidence of [1] and [2]) (Moore 2015, 50). All of this seems right; but are these conditions *sufficient* for peoplehood? Notably, the attachment to land is nowhere on this list. Presumably, it is something to which a people subsequently acquires a right, once it is ascertained that peoplehood in fact exists.

Before considering that issue, it may be asked which groups in fact would qualify for peoplehood, under these conditions. Moore initially lists two of them—inhabitants of existing states, and indigenous groups exercising (informal) self-government (Moore 2015, 53–54). Notably, national groups are not (initially) mentioned. Later on, Moore concedes that (cultural or ethno-)national groups may be included, to the extent that they have come to have a collective (political) identity. But this is an "explanation for why the group has developed a collective identity of a certain kind, not a justificatory argument for the group's exercise of jurisdictional rights" (Moore 2015, 61). Fair enough; but this does not assuage the worry that her concept is too "capacious" (her term), and will lead to justifying the claims of groups fighting over the same territories. We will return to this problem shortly.

The reason why Moore does not include territory in her initial definition of peoplehood is clear enough: it does not seem to play a role in political self-determination, and peoples are essentially self-determining entities. Cara Nine puts this point well: "The point of self-determination is that a group wants to exercise political power *over itself*" (Nine 2008, 49). There does not seem to be any intrinsic connection between territoriality and political self-determination. If a people is therefore defined politically, such a people will have no essential attachment to a place—such attachments may be instrumental to enabling self-determination, or not. Nationalists recognize this problem in attempting to formulate a conception of nationhood that is cultural—that is, based on beliefs about places that may be subjective, but nevertheless do establish a connection to such places that is not merely instrumental. Moore (rightly) rejects this nationalist view since it leads to inextricable conflicts between national groups with beliefs about the same or overlapping places. But this then leaves the connection between peoples and lands indeterminate—why exactly do peoples have rights to control over *particular* places (as opposed to *some* place in which to reside)?

Moore attempts to answer this question through espousing an "occupancy principle," namely, that peoples have the right to control the land on which they live, and that this involves borders "drawn *around* groups that meet the criteria of peoplehood" (Moore 2015, 118; see also 36). But why is this a matter of right, rather than just convenience? And, in cases where it is not convenient, as in the hard cases of territorial conflict, does viewing territorial claims as entitlements not exacerbate the very conflicts that she argues it should resolve? The concern, again, is that Moore's political view of territory is not so much wrong, as insufficient: the conditions for claiming territory are so broad as to include—and potentially justify—too many territorial claims, including those of national groups in conflict. What is needed is a *non*-political criterion (not necessarily a *pre*-political criterion, as nationalists such as Meisels or Miller might claim) for fixing territorial claims. Regarding these claims as political is too indeterminate; regarding them as *rights* is even worse.

We are back to the difference between a democratic theory of territory—that is, as determined by a democratic procedure—and a political theory of territory. While Moore rightly rejects plebiscitary democracy as unhelpful in resolving territorial claims, she merely substitutes a substantive view of democracy as providing the justification for claims by self-determining peoples. But the ideas of plebiscites and referenda were developed precisely to resolve conflicts between rival self-determining peoples claiming the same territories. Either way, whether the theory of territory is referred to as democratic or as political, something more is needed. In discussing hard cases (Ireland, Kashmir, Kurdistan—but not Israel/Palestine), Moore advocates the use of power-sharing institutions to resolve such conflicts (Moore 2015, 123–24). This is a helpful move that I have also endorsed in certain instances (Dahbour 2016). Nevertheless, as she admits, this is not a matter of (ethical) principle (i.e., justice), but of "institutional design." Why is it mandated by the "occupancy principle" (as I argue it is by a certain construal of self-determination)? In the end, the "political" theory of territory devolves into a democratic theory, a theory that begs crucial questions about how to resolve conflicting territorial claims between rival peoples seeking self-determination, *in the same place.*

This conclusion seems justified by the failure of the (purely) political theory to establish a close connection between peoplehood and territoriality. Since peoples are defined, and territorial rights are justified, without recourse to any intrinsic connection between peoples and territories—other than the incidental occupation of certain lands—there cannot be any principled right to maintain the connection. If another suitable geographical venue for collective self-determination is found, what justification do peoples have for insisting on continued control over a particular place? Whether through partition, "population transfer" (i.e., "ethnic cleansing"), or mass emigration, as long as there are some means—and some place—for a people's collective will-formation to be expressed, there does not seem to be any principled reason why, on the basis of the political theory of territory, one place is any better than another.

It is instructive to note where in the theory attachment to land does return—since this is after all the point of territorial rights in the first place. It is as a right to control, not territory, but *resources*, that territoriality manifests itself again: "the idea of collective self-determination offers a limited and defeasible right to control the rules governing the acquisition, transfer, and use of natural resources" (Moore 2015, 174). In discussion of different views of resource rights, Moore rejects both total control, and no control, of resources by making a distinction between the control of, and benefits from, resources. Furthermore, in certain cases, she argues that her theory "might involve eschewing development of a particular piece of land, perhaps because . . . development of the resource would be ecologically damaging, or because the land itself is viewed as sacred territory" (Moore 2015, 175). While this concession is laudable, it does not seem to be theoretically mandated but is simply a way of accommodating diverse views of resource use.

In more recent work, Moore has made the connection of peoples to resources, using the concept of self-determination, a closer one. Here is how she puts it: "jurisdictional authority over resources is an important dimension of collective self-determination. If people lack control over the land that they live on . . . then, to that extent, they lack robust forms of collective self-determination" (Moore 2019, 62). Still, this is not a conceptual connection, since self-determination is defined independently of attachment to, or control over, land. In addition, resources are not the same as land, since resources are sometimes fungible (or substitutable), and therefore less tied to peoples' intrinsic understanding of themselves. Finally, territory has different connotations from resources, inasmuch as territories (as discussed in Chapter 3) can become "homelands" for peoples in a constitutive sense.

The main point is that, as long as the theory is derived *only* from political principles (i.e., self-determination), there does not seem to be any essential connection or attachment between (self-determining) peoples and particular places or territories. Given the needs of democratic institution-building and maintenance, any number of places will do. The problem here seems to be that, as long as peoples are thought to be entities that can be defined without explicit and necessary reference to territory, then they may later on be viewed as potentially "floating free" of any particular locale. The right to territory therefore remains a right to a territory somewhere, but not to any particular territory.

Yet, even the understanding of political self-determination that underlies this political theory of territory requires specification with regard to some territorial entity. Self-determination does not, and cannot, simply mean the right to *express* the intention to amalgamate collectively, as a people. It must also mean the capacity to *embody* this intention concretely and materially in a *place*. This is why peoples and territories are conceptually linked; the very idea of a people without some territorial setting becomes nothing more than the conception of a group of individuals who choose (perhaps on contractualist grounds) to associate together.

To put it somewhat differently, (political) self-determination is not an "opportunity concept," it is an "exercise concept" (Taylor 1979). The right must be acted upon for it to be claimed, and to act on it is to exercise intentions to enact practices, policies, and other measures for a people, in a place. Without a place, there is no ability to enact anything, but only the (abstract) wish to do so. Territories are therefore essential to becoming a people, just as peoples are the only agents who can specify a territory (as opposed to an abstract space). What remains, however, is to understand the purposes for which territories can be legitimately claimed, and territorial rights justified.

An Ecological Concept of Territorial Rights

It is time to draw the various strands of the argument concerning territorial rights together. The specification of territory as the use, and therefore appropriation, of land by peoples, as was made by Kolers, is the first strand. The identification of the importance of territory for peoples who seek collective self-determination in a place, as argued by Moore, is a second strand. These are tied together by what I have called the conceptual argument—namely, that territories and peoples are mutually defining. Territories without peoples are land areas, but not discernible as anything with greater significance for human life. Peoples without territories are groups of persons who may aggregate for various reasons, but without forming anything that can seek political self-determination. Put together, we can say that peoples may claim the territories that (partly) *define them*, so that they may exercise their collective right of self-determination.

There is, however, a third strand in the argument for territorial rights, and it was mentioned at the beginning of this chapter: it concerns the ultimate purpose of territorial rights. This is the maintenance of natural goods and services, located within, and limited to, distinctive ecosystems—and which are necessary to sustain the life of the people resident in that ecosystem. This strand concerns the *stewardship* of these natural goods and services. Stewardship, then, pertains to the specifically ecological dimension of territorial rights, a dimension that, until quite recently, was ignored in justifications of such rights. This dimension is distinct both from proprietary conceptions of territory that equate concern for "natural resources" within a territory with private property rights over them (as discussed in Chapter 3), and from political conceptions that fail to adequately specify the purposes of claiming territory on the basis of peoples' right to self-determination.

Before examining the inability of such theories to justify the ecological stewardship of territories, the ecological concept of territorial rights advocated here can be stated as follows:

> *Peoples inhabiting territories comprising distinct ecosystems have the right to jurisdictional authority over those territories sufficient to maintain the goods and services of those ecosystems.*

The general justification for this jurisdictional authority of peoples over the territories they inhabit is the idea that rights to territories imply corresponding duties to the lands and the natural systems that they contain over which rights are claimed. This is an instance of the general philosophical notion of the correlativity of rights and duties, mentioned earlier. Peoples have duties to maintain the ecosystem goods and services without which they cannot sustain their habitation of the territory. Discharging this responsibility gives them the corresponding right to that territory.

It is noteworthy that, in recent years, conceptions of territorial rights have been developed that increasingly focus on this stewardship dimension; two such conceptions will be discussed next (Espejo 2020; Nine 2022). It is perhaps unsurprising that earlier property-oriented theorists of territory did not emphasize this ecological aspect. But there is one sense in which proprietary theories can be seen to address the environmental dimension of territory. This is in their recognition, as Cara Nine writes, that "jurisdictional authority" is essential for "solving important [collective] coordination problems" (Nine 2008, 49). But this authority need only have the "capacity to adequately solve the coordination problem to the extent that it secures the objects of its members' basic needs." No particular policies—environmentalist or otherwise—are mandated, because "there are many different adequate solutions to many different coordination problems [and] we cannot hold an authority to any particular solution."

The "political theory" of territory has a similar problem to the proprietary one: it is not sufficiently concrete in specifying preservation and maintenance of natural goods as essential conditions for a territorial right. Though Margaret Moore, for instance, has elsewhere written about the need to maintain "natural resources" (as mentioned earlier), territorial rights per se are not conditional on any particular good or goal to which a people claiming such rights must attend. When Moore discusses territorial conflicts, there is an indeterminacy in her treatments that belies the vagueness of her conception of how peoples are situated in relation to specific territories.

For instance, she uses the case of the Amazon River Basin, in which

> there is a direct conflict between the claims of the urban poor, who argue that their poverty should be addressed through, among other things, exploitation

> of the natural resources of the Amazon, and the claims of Amazonian indigenous people . . . who tend to resist attempts to exploit these resources.
> *(Moore 2015, 174)*

She concludes by writing that "we are sympathetic to both kinds of claim." Perhaps so, but how to adjudicate them? Though she does not return to this case, she makes two points: the indigenous people make the claim of self-determination (which the urban populations presumably do not), but this does not necessarily give them the right to the benefits or proceeds from exploitation of the area's resources.

These are important points to make, but they are not well integrated into the theory. This is because there is no substantive good that is posited as a feature of attachment to territory. Apart from needing an attachment somewhere, peoples need not adopt any particular stance or way of life, in order to claim occupancy rights. This is a recipe for conflict, since—other than affirming the importance of *some* territory—Moore does not give us a principled way to adjudicate these conflicts. This is probably the result of her view that such attachments are variable and, in fact, not related to anything essential in the lives of groups. On the one hand, the attachments are not primarily material, but "spiritual": "The attachment to . . . particular land can be explained *only* by appealing to non-instrumental . . . e.g., symbolic or cultural . . . values" (Moore 2015, 180, italics added). On the other hand, the necessary functions that land plays in the production and reproduction of life happens as a matter of course stemming from simple occupancy: "Transforming the land and being transformed by it will happen *naturally* through living on the land, using it as a source of food or shelter or water and interacting with the environment" (Moore 2015, 86, italics added).

Here we have the root of the problem: a positing of a nature-culture dichotomy in which material life is "naturally" produced, while attachment to *where* it is produced is "culturally" mediated. The dichotomy relies on a misreading of what "culture" (and also, "nature") is—neither is meaningfully separated from the other, and without an understanding of this, the attachment to land cannot be understood. As Raymond Williams reminds us in his book, *Keywords*, "culture" originated from its use in terms such as "crop culture" or "animal culture." It was later used to refer to particular ways of life; and finally, as the social sciences employed the term, to "material production" *or* "symbolic systems." Williams concludes that this "often confuses but even more conceals the central question of the relations between 'material' and 'symbolic' production . . . which have always to be related rather than contrasted" (Williams 1983, 87, 90–91).

In the domain of territoriality, the cultural or symbolic meanings attached to land are inseparable from the uses to which the land is put—usually having to do with the material production and reproduction of life *of a specific sort*. It is this specificity, a particular kind of socioecological system, that generates the claim to territory on the part of a people—but *also* defines who that people is. The definition is *internally related* to use of the territory, just as that use is internally related to the people that are engaged in that use. This is what is meant by a conceptual connection between people and territory.

This connection is *necessarily* embodied in sustainable uses of natural goods and services. This is why ecological stewardship is a necessary condition for a territorial rights-claim to be justified. Non-sustainable use of environmental goods, for instance, increased commodity production, maximal resource exploitation, high-yield agricultural cultivation—in short, extractivist practices—are inconsistent with legitimate claims to territory, *whether or not they are perceived as instrumental to the political self-determination of a group*. The reason is this: such groups cannot constitute peoples, because they are not *necessarily* attached to any particular territory. Such extractive uses can be practiced on any number or type of ecosystems; there is nothing specific about them that is related to particular environments.

The relation to territory that a people can have is defined by purposes and activities that are essentially place-oriented. If a population is largely engaged in other types of purposes and activities—for example, capital accumulation, military conquest, settler colonization, nonrenewable resource exploitation—then this would disqualify it from being regarded as a people. Such a population may constitute a national group, a political movement, the citizenry of a state, or a colonial settler group; but they do not constitute a people. Only a sustainable relation to a territory qualifies as such a constituting relation. This type of relation—one that involves ensuring ecosystem resilience within a territory—is what makes a disparate group of persons into a people.

Recently, philosophers have begun to notice this relation between peoples and territories. At least two exemplary studies have been published in the last few years—Pauline Ochoa Espejo's *On Borders* (2020), and Cara Nine's more recent book, *Sharing Territories* (2022). Their works seek to remedy what had been an "ecological blindspot" in the debate on territorial rights (Dahbour 2016). As the title indicates, Espejo's work is not a study of territorial rights per se, but rather is about justifying borders, though she has some general things to say about territoriality (Espejo 2020, 183). Nine's book is more centrally concerned with territory; but it may be doubted whether a clearly articulated right—at least if it is considered a collective right—is forthcoming from her study.

Despite these shortcomings, both works contribute significantly to the concept of an ecologically oriented theory of territorial rights. Espejo gives a strong argument for the importance of place-based rights and duties in political life. Furthermore, she argues vigorously (and correctly, in my view) for the validity of borders—though it is not always clear what these borders are demarcating (certainly not always states). Nine, in her recent work, has decisively broken with her earlier proprietary approach (discussed in Chapter 3). She now advocates a theory of territorial rights that emphasizes sustainability considerations and collective self-determination of what she calls "foundational territories." All of this is laudable and helpful in delineating what is essential in order to justify territorial rights.

One notable characteristic of both works is their emphasis on spatial metaphors for place-based (or territorial) rights. In Espejo's case, the metaphor of a watershed is used, though she is quick to add that it is (only) metaphorical, and not exclusive of other place-based claims (Espejo 2020, 180). Nevertheless, she does claim a unique status ("compactness") for watersheds that would seem to give them priority in relation to other kinds of territories (Espejo 2020, 176). Nine similarly emphasizes a water-based marker of place—in her case, rivers or "river catchment areas" (Nine 2022, 15–16). It may be wondered whether there is a "misplaced concreteness" in these spatial metaphors. In other parts of the works, reference is made instead to "social-ecological systems" or to ecosystems generally (Espejo 2020, 187; Nine 2022, 74). Perhaps not too much should be made of these stand-ins for more general features of ecosystems or socioecological systems.

There is one possible consequence of this emphasis, however; it is occasioned by the following question: In the articulation of border or territorial rights, where are the *peoples* who would supposedly claim such rights? Largely absent from both of these works, it seems. Not only peoples in general but also peoples' movements—politically mobilized efforts to make territorial rights-claims (such as those discussed in Chapters 1 and 2 of this book)—are also missing. References are made, in the case of Espejo, to various situations that may (or may not) be clarified by the spatial metaphors that she uses. For Nine, however, it seems to be mainly individuals (though not now proprietary individuals) that are the key claimants, at least theoretically, of territorial rights (though they must associate together in some way into foundational territories).

In her articulation of borders as indicative of "place-specific obligations," Espejo criticizes others who use the concept of the commons as a basis for non-state-oriented mutual obligations and institutions (Espejo 2020, 148–49). This would seem to be as close as she comes to discussing collectivities who might make claims or assert rights. Yet, later on, she seems to indicate that commons can, under certain circumstances, act as instances of "place-specific cooperation"; undoubtedly, the circumstance is "clear boundaries" (Espejo 2020, 156–57). Throughout, there is an odd lack of engagement with actual political struggles over boundaries and the goods contained in them. This is not for lack of examples, but rather that these examples are not systematically theorized in terms of rights-claims.

While the book has made other contributions (see earlier), it seems not to grapple with political conflict, despite acknowledging that it can center on "different management visions" of land and water use (Espejo 2020, 182–83). Most concerning is the lack of clear commitment to environmentally sustainable uses over other ones. Certainly, Espejo advocates borders that "run along . . . social-ecological systems" (Espejo 2020, 187). But when she indicates the "characteristic features" that could be used to mark off one region from another, uses such as livestock grazing, plantation agriculture, and coal mining are mentioned (Espejo 2020, 189). So borders that may designate ecological unsustainable usage may qualify as legitimate on her account—or at least are not prima facie disqualified by the basic theory of borders.

On this point, Nine's work is much more explicit. She asserts a "sustainability proviso" that is presumably a general condition for claiming territorial rights: "a system of rights over resources must ensure sustainable management of ecosystems and resources to support the capacity of current or future generations to live together sociably in the area" (Nine 2022, 74). As she puts it later in the book, any claim to self-determination must be constrained by this proviso, among others (Nine 2022, 252). So far, so good. The problems, however, arise with the more basic conception of territorial rights that this proviso modifies. Again, as with Espejo, the importance of territoriality (or place, in Espejo's theory) is asserted in the absence of a robust conception of peoplehood. In Nine's case, this omission seems to be intentional.

Nine stigmatizes theories that include peoplehood as a condition of territorial rights as attachment, or relationship-based, theories (Nine 2022, 81, 84–85). She sees a number of problems with such an account: it over-valorizes social identities (such as national identity); it is statist in orientation; and, above all, it is unable to deal with the many instances of territorial *overlap* (Nine 2022, 138–40). While some of this may be a problem for theories that focus *exclusively* on peoplehood as a basis for territorial rights (e.g., nationalist theories such as those of Meisels or Miller, and arguably political theories such as Moore's), it should not be a problem when peoples and territories are co-determining, or conceptually linked, as I have argued earlier. More likely, the relative absence of consideration of the role of peoples in defining the territories upon which they depend for a sustainable livelihood will be the problem.

Nine seeks to avoid this by establishing a basic "use-right" to resources that is foundational for rights to territory (Nine 2022, 77). But notably, this use-right seems to be a right of *individuals* to use a resource (such as the water from a river). Recognizing that this is problematic in various ways, Nine then seeks to amalgamate such use-rights into "foundational territories" that share a place, such as a "river catchment area" (Nine 2022, 153ff). Such territories may lie across existing states, and require specific forms of political authority to deal with the issues of overlapping jurisdiction that go with such places. Nine believes that an advantage of her theory is that it can deal with territories in which resources need to be shared between competing users (Nine 2022, 239).

But can it resolve such conflicts absent a theory that settles *which* users (what peoples) get priority? Nine uses the example of Los Angeles as a city that gets (much of) its water from the Colorado River (Nine 2022, 51). She uses this example without apparently considering that such use is: [1] unsustainable and [2] unjust inasmuch as it is the result of political power overriding native and regional use of the river. To use Espejo's theory for a moment, California is *not* a part of the Colorado River's watershed (i.e., ecosystem). Should that make any difference? What are the limits to "sharing," on Nine's theory? In the absence of establishing a stronger connection between territories and peoples, the answer is apparently that there is no limit. But how then can such a theory abide by the "sustainability proviso," that she asserts early on? By attempting to develop a theory that can satisfy *all* users, Nine has apparently made it much harder for anyone (any people) to assert territorial *rights*, even in cases where such assertions are founded on close connections between peoples and territories.

This is, by the way, not how the conception of common property institutions, that both Espejo and Nine reference, works. Common property is a matter of a body of sustainable users determining the limits on resource use, and consequently limiting—in fact, *prohibiting*—use by others (Ostrom 1990). The ecological theory of territory advocated here is a means of granting such users this political authority *by right*. Sharing does have its limits.

To summarize, the ecological theory of territorial rights consists of three elements—territoriality, peoplehood, and sustainability. The first two elements are combined into an ethical principle of territorial rights based on a conceptual argument about the codetermination of territories and peoples. On the one hand, territory is defined by the relation of land to a particular people who inhabit it. On the other hand, this people is entitled to rule over the territory to the extent that it is necessary for that people's livelihood. The ethical principle has a proviso, or side-constraint—that the right is forfeit if the territory is not maintained in an ecologically sustainable condition across several generations (as determined by local cultural traditions concerning inheritance). These elements are the basic building blocks of the theory. But they do not answer questions that arise in special circumstances. Four such circumstances can be considered here briefly (some have been, or will be, discussed in other chapters): [1] cases of culturally distinct peoples competing for resources within the same ecosystem; [2] cases in which this is the case, plus the additional element of outsiders settling in the area (i.e., settler colonialism); [3] cases of peoples who were forcibly excluded from an area, who wish to resettle it; and [4] cases of unsettled or lightly settled areas, where landless outsiders are seeking land.

In the case of culturally distinct peoples in competition (e.g., such instances in recent history as the partition of India or the dissolution of Yugoslavia), the question is whether such peoples are able to cooperate in living within the ecosystem, despite cultural differences. Narratives and theories of power-sharing, in areas from Ireland to Israel-Palestine, suggest the possibility of success in this regard (Cordell, O'Leary, and Wolff 2016). In any case, the matter of cultural difference does not seem to establish any necessary priority of rights. Historically, political movements that are not aligned with either group have often existed (as in Punjab in the 1940s, and in Bosnia in the 1990s). These have *ethical* priority, though the balance of political forces may not always allow them to prevail.

In the settler-colonial case, the additional element of outsider settlement complicates, but does not fundamentally change, this situation. Indigenous peoples, already settled and using natural goods and services for their livelihood, take priority in settlement. This does not mean that outsiders (see the fourth case) may not be invited to settle; but the decision is one that must be made in a procedurally fair way by the already existing inhabitants. Obviously, in the case of settler colonialism (e.g., early modern North America, Palestine in the twentieth century), the presence of colonial powers often makes this impossible. Again, the ethical priority lies with the inhabitants; settlement must not be coerced.

Sometimes peoples have been forcibly excluded from ecosystems in which their ancestors lived (again, the European settlement of North America provides multiple cases of this). Do they have a "right of return"? Such a right may be invoked in the settler-colonial case, genuinely or disingenuously. But a legitimate case would involve claims to *re*settlement within a generation or so. There must be a memory trace held by living descendants of individual family members from the area; otherwise, ideological invocations of purported, but nondemonstrable, ties to regions could become ubiquitous.

Finally, can landless peoples (e.g., the Landless Workers Movement in Brazil) claim rights of settlement in underutilized areas (that is, from their point of view)? The primary consideration that comes into play here is the following: Would settlement degrade the ecosystem to such an extent that it could not support current inhabitants *in their present form of livelihood*? If not, then settlement is allowable (without prejudice as to the cultural origins of the settlers). If so, then settlement must be restricted to allow the continuation of current inhabitants' livelihoods. Obviously, the most problematic cases are those in which settlers seek to radically change the environments they are trying to settle (e.g., farming in the Amazon rain forest). Other cases, however, even if they involve radical disjunctures between current and settler populations, may be allowable (other considerations being equal) (e.g., landless farmers from South Asia settling depopulated farmland in North America).

These observations are relatively ad hoc, since the main point here is to establish fundamental principles that can be used to adjudicate these and other cases in the future. These principles eventuate in an ecological right to territory, justified as a right of peoples inhabitating ecosystems to maintain both their own livelihoods in an existent socioecological system, and the overall ecological sustainability of the system for other living creatures as well. Such a right, again, is more than a property right; it is a political entitlement that involves collective forms of social organization, and the rights and duties that go along with such forms. Paramount here, in terms of realizing such territorial rights, is a claim to sovereignty over the ecosystem in question.

Why Sovereignty Is Necessary for Territorial Rights

The political institutions and regimes necessary to sustain ecosystems remain undertheorized. Are ecosystem-based territories best governed by states, common property institutions, or some combination of them? Does the theory of territorial rights defended here support state authority, or undermine it? In both instances of common property institutions and of sovereign territorial states, territoriality is governed through regimes that seek control over bounded spaces. But there are important differences. For peoples, inasmuch as they are rooted in a relation to an environment, that environment will usually *not* map onto state boundaries, which are artifacts of political struggles at a larger scale. That means that commons institutions will be necessary to govern territories in order to ensure the ecological sustainability of those territories. States will sometimes act to do this; but more often, they will seek to exploit the resources of ecosystems for their own purposes, or those of the social elites that they serve.

Sometimes, states can be valuable means of protecting the integrity of territories from other states, or from market institutions that seek to capitalize environmental goods and services. But extractive practices are often perpetrated by local states themselves. In such cases, peoples must mobilize *against* the state in some way, and commons institutions transform themselves into political vehicles for popular resistance. It is important that the sovereignty principle be understood in such a way that it can be invoked by these institutions against governments that may seek to justify their actions by claiming exclusive sovereignty rights for themselves. Rethinking the principle of political sovereignty for the age of environmental crisis, such that it can be used for the protection of ecological sustainability, rather than to legitimate extractivism by states, is the purpose of the next two chapters.

References

Agnew, John. 2009. *Globalization and Sovereignty.* Lanham, MD: Rowman & Littlefield.

Beitz, Charles. 1999. *Political Theory and International Relations.* Revised ed. Princeton, NJ: Princeton University Press.

Blaut, J. M. 1979. "Some Principles of Ethnogeography." In *Philosophy in Geography*, edited by S. Gale and G. Olsson. Dordrecht: Reidel.

Breuilly, John. 1982. *Nationalism and the State.* Chicago, IL: University of Chicago Press.

Canovan, Margaret. 2005. *The People.* Cambridge: Polity Press.

Cordell, Karl, Brendan O'Leary, and Stefan Wolff, eds. 2016. *The Israeli-Palestinian Conflict: The Politics of Stalemate* (a special issue of Ethnopolitics 15: 4). Routledge.
Dahbour, Omar, ed. 1997. "Philosophical Perspectives on National Identity." Special issue of *Philosophical Forum* 28 (1–2, Fall-Winter 1996–97).
Dahbour, Omar, ed. 1998. "The Incoherence of Liberal Nationalism." In *Memory, History, Critique: European Identity at the Millenium*, edited by Frank Brinkhuis and Sascha Talmor. Utrecht: ISSEI.
Dahbour, Omar, ed. 2003. *Illusion of the Peoples: A Critique of National Self-Determination.* Lanham, MD: Lexington Books.
Dahbour, Omar, ed. 2013. *Self-Determination Without Nationalism: A Theory of Postnational Sovereignty.* Philadelphia, PA: Temple University Press.
Dahbour, Omar, ed. 2016. "Self-Determination and Power-Sharing in Israel/Palestine." *Ethnopolitics* 15: 393–407.
Deudney, Daniel. 1996. "Ground Identity: Nature, Place, and Space in Nationalism." In *The Return of Culture and Identity to IR Theory*, edited by Y. Lapid and F. Kratochwil. Boulder, CO: Lynne Rienner.
Elden, Stuart. 2010. "Land, Terrain, Territory." *Progress in Human Geography* 34 (6): 799–817.
Espejo, Pauline Ochoa. 2020. *On Borders: Territories, Legitimacy, and the Rights of Place.* Oxford: Oxford University Press.
Horowitz, Donald. 1985. *Ethnic Groups in Conflict.* Berkeley: University of California Press.
Kolers, Avery. 2009. *Land, Conflict, and Justice: A Political Theory of Territory.* Cambridge: Cambridge University Press.
Kolers, Avery. 2012. "Justice, Territory, and Natural Resources." *Political Studies* 60: 269–86.
McCarthy, James. 2007. "Territoriality." In *Encyclopedia of Governance*, edited by Mark Bevir. London: Sage Publications.
Meisels, Tamar. 2005. *Territorial Rights.* Dordrecht: Springer.
Miller, David. 2012. "Territorial Rights: Concept and Justification." *Political Studies* 60: 252–68.
Moore, Margaret. 2015. *A Political Theory of Territory.* Oxford: Oxford University Press.
Moore, Margaret. 2019. *Who Should Own Natural Resources?* Cambridge: Polity Press.
Nine, Cara. 2008. "A Lockean Theory of Territory." *Political Studies* 56: 148–65.
Nine, Cara. 2012. *Global Justice and Territory.* Oxford: Oxford University Press.
Nine, Cara. 2022. *Sharing Territories: Overlapping Self-Determination and Resource Rights.* Oxford: Oxford University Press.
Ostrom, Elinor. 1990. *Governing the Commons: The Evolution of Institutions for Collective Action.* Cambridge: Cambridge University Press.
Pateman, Carol. 1985. *The Problem of Political Obligation: A Critique of Liberal Theory.* Berkeley: University of California Press.
Sack, Robert. 1983. "Human Territoriality: A Theory." *Annals of the Association of American Geographers* 73 (1): 55–74.
Sassen, Saskia. 2013. "When Territory Deborders Territoriality." *Territory, Politics, Governance* 1 (2): 21–45.
Sheppard, Eric, and Robert McMaster. 2004. *Scale and Geographic Inquiry: Nature, Society, and Method.* Oxford: Blackwell.
Stilz, Anna. 2019. *Territorial Sovereignty: A Philosophical Exploration.* Oxford: Oxford University Press.
Taylor, Charles. 1979. "What's Wrong with Negative Liberty?" In *The Idea of Freedom*, edited by Alan Ryan. Oxford: Oxford University Press.

Whelan, Frederick. 1983. "Prologue: Democratic Theory and the Boundary Problem." In *Liberal Democracy: Nomos XXV*, edited by J. R. Pennock and J. W. Chapman. New York: New York University Press.
Williams, Raymond. 1983. *Keywords: A Vocabulary of Culture and Society.* Revised ed. New York: Oxford University Press.
Yack, Bernard. 2001. "Popular Sovereignty and Nationalism." *Political Theory* 29.

PART III

Sovereignty Claims

5

THE REVIVAL OF POLITICAL SOVEREIGNTY

Sovereignty Redux

A central argument of this book is that opposition to extractivism and its environmental effects must become political, if it is to be effective. But what does that mean? Above all, I contend, it must involve sovereignty claims by anti-extractivist forces that assert ultimate authority over the socioecological systems that require protection. But that means that such sovereignty claims must be legitimate in the sense that their assertions of authority are justifiable, and that those who assert them are the political actors who are rightfully entitled to do so. These two aspects of sovereignty—the justifiability of sovereign authority in general and the legitimacy of the specific authorities who could use sovereignty claims to oppose extractivism—are the subjects of this and the next chapter, respectively.

First, the idea that sovereignty remains an important feature of political action, justifiable on ethical grounds, will be discussed in this chapter. Among other things, this will require that we establish a justification of sovereignty that differs from the one usually given—that is, the "absolutist" justification that is found in early modern political theorists such as Bodin and Hobbes, and is often still regarded as canonical by recent philosophers. Second, the idea that anti-extractivist human communities such as agrarian peoples or indigenous groups are the legitimate claimants of sovereignty will require separating sovereignty from *state* sovereignty, something that also goes against the grain of much contemporary political theorizing about such matters (this issue will be taken up in Chapter 6). Together, this will yield a full statement of the ecosovereignty principle, given toward the end of the next chapter.

DOI: 10.4324/9781003568155-9

One thing that unites different versions of the argument for political sovereignty is that the doctrine is necessary, in fact essential, for the legitimate exercise of political power. This idea has made a perhaps surprising reappearance in the political thought of the early twenty-first century. After the heyday of conceptions of global governance, democracy, and justice in the decade following the end of the Cold War, the costs and dangers of minimizing local community and its claims to sovereignty have become manifest. While by no means completely eclipsed by locally oriented political theory that valorizes sovereignty, globalizing ideas of community are no longer regarded as the obvious context for legitimate political action. As we will see later in this chapter, such views are now regarded as contentious and controversial, perhaps flawed, in a way that was not the case until quite recently. Instead, the idea that geographical place and locality are essential contexts for political action and that consequently sovereignty remains an important political norm is now an increasingly prevalent view. This chapter will examine why this is so—why sovereignty again seems a necessary concept for solving myriad problems, including the environmental crisis.

One aspect of why sovereignty is essential for dealing with extractivist and ecocidal practices and policies concerns the limitations of territorial rights. While it is certainly the case that, as argued in the last chapter, territorial rights are best founded upon a concept of ecosystem integrity and sustainability, this may not be sufficient to curtail extractivism in many instances. The argument given was for a conception of territorial rights that could override proprietary conceptions of territory, among others. This means that private, business, and corporate practices aimed at extracting natural value from ecosystems would be subject to limitation and cessation on the basis of the territorial rights of ecosystem-based peoples who seek to maintain these ecosystems. This would be a tremendous advance over rights regimes that prioritize private property rights to natural goods and services.

But what about governmental policies designed to do the same thing? In many traditional legal systems, states retain (sovereignty) rights to override property rights—whether to private or communal property—on the basis of "reasons of state." These considerations may involve the securing of goods such as public transport (routes) or water reserves. But they may also aim to favor or privilege private or public extraction of agricultural, pastoral, or forest lands or resources. If states maintain such sovereignty regimes, they can potentially abet or institute extractivist initiatives, overriding territorial rights.

A sovereignty regime, as we will see, is about having authority that potentially overrides all other authorities, rights, and claims. Without such *ultimate* authority, communities and peoples risk subsumption under others with antithetical goals or priorities. This is the true meaning and significance of sovereignty, as it exists above and beyond territorial claims and rights.

Once the general argument for the continuing importance of sovereignty is complete, we will turn to consider recent changes in the sovereignty doctrine that render it useful for opposition to extractivism, among other things. The way in which this will become clear is in the connections that have been established between sovereignty and the idea of (political) self-determination, particularly in the postcolonial context. It is there that the importance of sovereignty in opposing various extractivist initiatives, as they are justified by neocolonial, neoliberal, and other ideologies, including that of "sustainable development," is most evident. But first, we must clarify the basis for a coherent and legitimate sovereignty doctrine, separating it from received notions of sovereignty in political philosophy.

The Absolutist Conception of Sovereignty

The original conception of *modern* sovereignty is often taken to be equivalent to the idea as such. The principal basis of this confusion lies in taking exponents of the concept, especially Jean Bodin and Thomas Hobbes, as giving a statement of the basic concept, rather than a particular interpretation of it. In their conceptions, the role of the state—and even more so, the ruler of the state—played an essential role. Sovereignty was regarded, not fundamentally as a principle applicable to a number of actors and situations so much as a faculty or feature of certain persons. This "misplaced concreteness" resulted from attributing (sovereign) authority in its entirety to one institution (a state or government) or agent (a ruler or monarch). This attribution, in turn, led to the conflation of final with ultimate authority. If the monarch, for instance, was sovereign, this meant that they were enabled to make a final decision or determination (e.g., an executive decision or a diplomatic agreement), *and that the decision or determination was necessarily authoritative*. But this is to conflate a *final* authority with an *ultimate* one. Since sovereignty is a principle, it can be misused or misapplied. An ultimate judgment about something may differ from the decision that was taken about it, if the decision overlooked or violated some important consideration (Benhabib 2009, 93). The consequence of this distinction is that sovereignty is not a function of a person or an office, but a principle. Sovereignty, in other words, is the *idea* of (ultimate) authority itself; it is not synonymous with someone (or some institution) that exercises authority.

In early modern formulations of sovereignty, this distinction did not come out clearly. Furthermore, arising in the face of ongoing continent-wide wars in seventeenth-century Europe, sovereignty did not, as in the concept articulated by Bodin, seem to require a philosophical justification (Bodin 1992). The need for a basis upon which to build civil and interstate peace, in the face of interminable warfare, seemed to mandate the concept. Sovereignty provided a juridical basis for peacemaking, both within and between states. But in order not to be regarded as simply an ad hoc means of conflict resolution, sovereignty needed a more thorough justification, against the continuing claims of would-be continental or global hegemons such as the papacy, the Holy Roman Empire, and the more powerful royal families in Europe (e.g., the Bourbons or the Hapsburgs).

The contractarian justification for sovereignty given, for instance, by Hobbes, which has been the most widely recognized one since this period, seemed to fit the bill. At the same time, such a justification also worked too well: it justified an absolutist conception of sovereignty that was ill-suited to the new iterations of the sovereignty doctrine from the eighteenth century onward. Only a conception that accords absolute and ultimate authority to a ruler, preferably a personal one, seems to follow from this justification.

How does the justification go and how did it go wrong? First, sovereignty, and political authority in general, is regarded as that state or condition in which freedom is absent; or, to put it differently, freedom is a condition in which there is an absence of authority. Freedom, in other words, is regarded negatively, as the absence of constraint or inhibition of choice or action. Second, an agreement between persons is the result of an arbitrary choice—that is, their decision to agree. This choice is the one enabling condition of the agreement; persons are free individuals, potentially self-sufficient, at least if they are not inhibited from securing the conditions for that freedom. In short, they are "possessive" individuals, capable of acting, and of taking what they may need (from the natural world) independently of any association with others. This act of possession, of which all are capable, is a free, pre-social capacity of all persons. Third, the immediate context of this agreement or compact is an "emergency" condition in which such individual freedom as persons naturally exhibit is threatened or contested. Absent such threats or dangers, there is no rationale for agreement. These three aspects of the argument need to be further considered in terms of their connection to a justification of sovereignty.

Freedom viewed as a condition of the absence of authority is, as just stated, generally regarded as negative, in the sense that to be free is to be free of others' impositions, prohibitions, limitations, and so on. Freedom is the absence, rather than the creation, of an optimal condition. Political authority, then, is that institution that takes away from the freedom of persons. Hobbes is the classic author who best articulated this view (Hobbes 1958). Authority, and sovereignty, may be necessary for social life—Hobbes certainly seemed to think so—but this does not make it a feature of human agency; quite the contrary. Except for the sovereign rulers themselves, authority detracts from, when it does not abolish outright, personal freedoms. In fact, this is required to a considerable degree for authority to function effectively, as Hobbes famously maintained.

If persons do give up their freedoms, they can only do so willingly by an explicit decision, opting out of a condition of freedom that has failed to provide them with even a modicum of peace or security. But this means that the need for a principle of sovereignty arises from an emergency, a breakdown, or a corruption—Jean-Jacques Rousseau called it a condition of slavery—that is episodic and disruptive (Rousseau 1978). Despite Rousseau's view that human dependency does emerge from the anthropological preconditions of society and the state, the development of conceptions of authority and sovereignty is not a way of making legitimate the governing institutions of already existent human communities. It is a means of addressing the absence of any stable social life, and the insecurities, threats, and dangers that result. Such a view suggests that sovereignty is none other than an arbitrary, though legitimate, act of creation.

It is tempting, therefore, to look for a more solid foundation upon which to base the notion of political authority, including sovereignty. Certain twentieth-century thinkers, influenced by early modern notions, have sought to portray sovereignty as rooted in the very nature of political action, much as divine sovereignty is rooted in the nature of divine will. This idea has found expression most famously in Carl Schmitt's work, as well as more recently in the scholarship of Jean Elshtain.

Both Schmitt and Elshtain view political sovereignty as a kind of secularized theology (Schmitt 1985, 36; Elshtain 2008, 91). However, as Elshtain admits, only if sovereignty is an undifferentiated and monolithic norm, justifying absolute authority, does a theological basis for it seem to be required (Elshtain 2008, 92). Furthermore, this "political theology" is controversial since, as Elshtain notes, Christian doctrine usually subordinates political sovereigns to the rule of (natural) law (Elshtain 2008, 115).

Schmitt, on the other hand, bases his claim that political sovereignty is a secularized theological idea on an understanding of the nature of law and legal decision-making. For him, judicial decisions are a template for political actions in general—they are decisionistic at the core. Schmitt's opening claim in *Political Theology*, "Sovereign is he who decides on the exception," equates political choices with legal decisions (Schmitt 1985, 5). Both exist in an underdetermined context in which judges (and rulers) must make their own interpretations and decisions, without any definitive guidance from precedents or principles.

The theological element here is metaphorical: sovereign political acts are not directly drawn from divine inspiration or Christian doctrine. Rather, the role of the sovereign in deciding legal cases (and making political decisions) is similar to that of God in theology: "The exception in jurisprudence is analogous to the miracle in theology" (Schmitt 1985, 36; see also Hirst 1987). What gives this comparison more cogency than it may at first appear to have is the very real problem of indeterminacy in legal interpretation (Scheuerman 1999, 232–33).

Schmitt's view of indeterminacy is that it requires a sovereign able to settle controversies and disputes about laws. Without this, no legal (or political) institutions can function:

> Whenever antagonisms appear within a state, every party wants the general good . . . . But sovereignty (and thus the state itself) resides in deciding this controversy, that is, in determining definitively what constitutes public order and security, in determining when they are disturbed, and so on.
>
> *(Schmitt 1985, 9)*

Two points are essential for Schmitt's argument that the sovereign is God-like to succeed. First, the role of principles and norms in adjudicating cases and conflicts must be minimized. If interpretation admits of better or worse results, based on adherence to agreed-upon norms, then the sovereign judge or ruler's discretion is less important. But, Schmitt asserts, "Like every other order, the legal order rests on a decision and not on a norm" (Schmitt 1985, 10). This is because sovereignty involves the "authority to suspend *valid* law," not just mistaken interpretations of decisions (Schmitt 1985, 9, italics added).

But this argument seems to be circular: if sovereignty did not have this power, then the state and the law would be more dependent on principles and norms, less subject to perhaps arbitrary decisions or rulings. Schmitt is asserting the absolute, quasi-divine nature of sovereignty in cases where the sovereign can *legally* override standing laws and institutions (as in the emergency dictatorship provision of the German constitution during the Weimar period in which Schmitt was writing) (Schmitt 2014; see also McCormick 1998).

Second, Schmitt has a point about the "God-like" character of sovereigns, *if* he is referring to the beginnings of institutions and laws: in *establishing* regimes, rules, and so on, there is certainly an important role for those who have the "exceptional" power to create such regimes. Schmitt was not the only thinker to make this point: Hannah Arendt also emphasized the role of arbitrariness in the *origins* of states. But for her, if constitutions (for example) originated in revolutions or wars, this was just to say that political life began with a *pre-political* (and often violent) moment of action. Politics was about what came afterward; it was not itself characterized by arbitrariness and violence, as Schmitt seems to be claiming (Arendt 1963, Introduction; see also Scheuerman 1997).

It is one thing for political actors to play an essentially "arbitrary" role in the *origin* of regimes; it is another for such a role to persist in the continuation and routinization of those regimes. For this difference to be overlooked, it is necessary to have a view of sovereignty that is essentially absolutist, even dictatorial. This is the view Schmitt in fact seems to have had; but it is certainly not a necessary one. Sovereignty is only quasi-divine, politics is only political theology, if the idea that political rule is a conditional matter of accommodation between divergent interests and views is rejected as an impossibility, in favor of an all-or-nothing conception of political authority.

This absolutist conception is derived from Hobbes; and it is no surprise that Schmitt, who espoused such a conception, has been referred to as the "Hobbes of the twentieth century" (George Schwab, in Schmitt 1985, xl). But, as will be discussed more later on, there is nothing essential in the sovereignty notion that requires its construal as an absolute, overriding other principles (or laws) in every case. This is not to take away from Schmitt's achievement in revitalizing the centrality of sovereignty, power, and politics in an era that has sought to instrumentalize it for other (e.g., economic) purposes. But even Elshtain, in some ways sympathetic to the idea of political theology that connects authority with divinity, admits, "At its best, sovereignty has taken the form of the constitutional, limited state that affords its citizens both security and liberty" (Elshtain 2008, 92).

Nonabsolutist Conceptions of Sovereignty

The history of the modern concept of sovereignty shows both the limitations of the concept and its necessity. If there is going to be political life, there needs to be authority of some kind. If there is authority, there will be an ultimate authority, even if only as an *idea* of the principles and considerations necessary for a political action to be legitimate. Sovereignty is the name for this ultimate authority, and it seems to be inherent in political action as such, at least potentially. Without such a notion, political life becomes not only conflictual often but also conflictual *always* since there is no principled or legitimate way to adjudicate and resolve conflicts. So one lesson to be learned from this history is the necessity of *some* concept of sovereignty.

The standard justification of the classical concept of sovereignty (that is, state sovereignty), one that relies on a decisionistic, not to say authoritarian, view of the concept, does not seem to work. Sovereignty cannot be justified simply by people's acceptance of a ruler as a means of establishing law and order. There are potentially other ways to achieve this result. Similarly, sovereignty is not justified because there needs to be some finality to political decision-making. Finality does not equal ultimacy, and has no normative status, but is simply a function of the urgency of political action at various times. Sovereignty must be ultimate, not final, authority, and this cannot be justified by the need for someone to make a decision in extraordinary situations.

A more general justification is needed in order to legitimate a principle of sovereignty. This justification must include several features. First, both the purposes and limitations of sovereignty must be reconsidered. Second, sovereignty must be adapted to contemporary conditions in terms of being of use in realizing norms of, for example, popular government, constitutional rights, social justice, and environmental sustainability that have become central to political life in the current period. The first matter involves regarding sovereignty not just as a matter of peace and security, nor as the necessary right of rulers or states. The second, as will be examined in the next chapter, is achieved by linking sovereignty with the emergent notion of self-determination, as it developed in twentieth-century colonial liberation movements and the postcolonial era that followed.

In the early modern conception of sovereignty found in Bodin and Hobbes, persons' liberty—as well as considerations of justice or welfare—were regarded as contradictory or extraneous to the concept. The possession of political authority that was the key to sovereignty was thought to include primarily the monopoly of coercion or force.

Yet, this need not—and may not always have been—the case. The protection of liberties and the provision of welfare—understood as necessary entailments of legitimate authority—could be a part of the sovereignty concept (Benhabib 2009, 81). James Turner Johnson has made the claim that an older, medieval notion of sovereignty that predates the ideas found in Bodin and Hobbes does incorporate such considerations. This older conception was first articulated, according to Johnson, by Augustine, and later elaborated by both Christian and Muslim medieval philosophers (including Aquinas). This older conception defines sovereignty "not in terms of the state and its territorial inviolability but in terms of the moral responsibility of the ruler for the common good of the people governed" (Johnson 2014, 2).

The view that Johnson propounds is controversial; he criticizes other historians of the sovereignty concept for viewing it only as a doctrine of the modern state. For Johnson, the older view is rooted in a conception of natural law that predates and overrides ideas about reasons-of-state. In his treatment of the two concepts of sovereignty, Johnson makes an argument about how the modern concept replaced the medieval one; but this need not concern us here. Even if his history is controversial, the importance of the view is that sovereignty admits of different definitions, including one that does *not* necessarily prioritize the rights of the state over those of other political actors.

The advantages that obtain from this older conception, according to Johnson, are various. First, it links "political order to justice and peace"; second, it subordinates power to "concern for the good of the community" and counterposes sovereignty to tyranny for that reason. It also sets limits on the use of armed force, establishes criteria for good government, and respects, to some degree, the "rights of the governed" (Johnson 2014, 107). Despite these aspects, the older conception of sovereignty was eventually modified and adapted to different uses in the period of the religious wars. As a result of that adaptation, the modern conception of rights of states to territory and independence came into being (Johnson 2014, 106).

Johnson's claims about the history of the sovereignty concept are perhaps ones that only a scholar of medieval philosophy can fully judge; nevertheless, they are suggestive for how we might think of sovereignty once the straightjacket of the classical, absolutist version is discarded. Nevertheless, there do seem to be two aspects of the modern idea of sovereignty that remain fundamental, yet were not made explicit before. These are the concepts of the territory and of the people. As Johnson puts it, the modern understanding of sovereignty required a "territory that may be defended" and a "people as having one fundamental right, that of self-defense" (Johnson 2014, 1).

As should be clear by now, the original modern conception of sovereignty denotes an idea of state rule that is as unlimited and unrestricted as possible. It includes both the idea that sovereignty is essentially authority that is unconstrained, and the idea that this authority is not only temporally final but also conceptually ultimate, in the sense that it is unlimited by other normative considerations. But this original modern conception is actually only one interpretation of the nature of sovereignty—and one that is ill-suited to contemporary conditions and considerations.

There are two ways in which sovereignty includes, in its very definition, a notion of the limitation, rather than the absolutization, of power. These ways apply to the two sides of sovereignty—internal authority (within the community) and external authority (in relation to other communities). On the one hand, the authority within a (political) community must be, in some fashion, the authority *of* that community (over itself). It cannot be the imposition of an alien power and still be, in any meaningful sense, legitimate within that community. So the issue is which element or elements of the community wield such authority. The classic modern answer is "the ruler" (e.g., a monarch); but this need not be the only—or even a very good—answer. A popular or constitutional or local authority could just as well wield sovereignty within a community. In fact, as we will see shortly, such an understanding of sovereignty is the outcome of the subsequent (post-1750) history of the concept.

On the other hand, external authority could not be maintained if power was absolutized by a hegemonic force exerted over all communities. For instance, if there were but one power over all (the world), no conception of sovereignty could be maintained—there would be nothing to which the sovereignty of a community could be related (i.e., maintained or asserted). In other words, without external autonomy from subordination to others, no sovereign could have *any* authority at all. It is a requirement of internal sovereign power that there not be only one sovereign power—but that those in a position of authority over a partial domain be *recognized* (by others) as sovereign *in that domain*.

In this sense, sovereignty is a doctrine of the limitation of power, not of its omnipotence. F. H. Hinsley, who has emphasized this point in his classic study of the doctrine, argues that this is a logical entailment of the notion, at least as applied to distinct political communities: "a state which claims to be free of limit and control within its community is bound in logic to concede the same freedom to other states in theirs" (Hinsley 1986, 158). If sovereignty justifies the notion that there is an absolute authority *in one place*, this means that there can be no greater authority over it emanating from some other place. Furthermore, sovereignty does not mean that one community or country or state has perfect freedom to act, but only that there is the "absence over and above the state of a superior authority" (Hinsley 1986, 226).

Sovereignty therefore means the following things. First, it is the *idea* of ultimate authority—but not as something held by any particular authority, whether a person or an institution. Second, sovereignty is an inherent part of political life in the sense that it is required for effective political action—but it has no super-political status, whether theological or moral. Third, sovereignty refers not only to the rights of the sovereign authority but also to their responsibility to ensure the welfare and liberty, as well as the security and independence of the community. Fourth, sovereignty is a conception of authority that specifies certain limits of all sovereignty authorities, both internally and externally.

The upshot of these aspects of sovereignty is twofold. Most importantly, sovereign authority is *not* equivalent to state authority. As Elshtain points out in her survey of ideas about sovereignty, authority is not even necessarily political—it *is* possible to have a divine sovereign, or even a moral agent as sovereign, at least in regard to certain aspects of existence (e.g., sacredness or moral rightness) (Elshtain 2008, xiii–xv). Furthermore, as we saw in Chapter 1, the concepts of producer or consumer (i.e., market) sovereignty are not metaphorical, but in some contexts real expressions of certain relations of power and authority. Of course, for our purposes, political sovereignty is what is at issue here. It is crucial, as noted earlier, that political sovereignty not be regarded as epiphenomenal on some other kind of legitimate authority (as in the idea of a political theology). Nevertheless, equally critical is the realization that political authority itself is not synonymous with *state* power. There are other political agencies—peoples, above all, constituted in various ways—that can assume this role.

In fact, the idea of "constitution" is critical here: as Elshtain has indicated, sovereignty is *best* understood as a constitutional notion of political power. It is important to consider more carefully what this means. Above all, it means that sovereignty—ultimate authority—does not mean *unlimited* authority. Rather, it means that power can be legitimately held by an authority which, while it can overrule other, subordinate authorities, holds this power subject to certain limits or constraints. This point has already been made in terms of external sovereignty, where there are geographical limits to sovereign power, by definition. But this is true of internal sovereignty as well.

While there have been a number of theories of constitutional authority, they all tend to agree on two points concerning the nature of sovereignty. First, constitutional authority involves a separation of powers, and the enshrinement of citizens' rights and duties, in key political documents and institutions. Second, sovereignty is not defined by a sovereign agent's identity, or even by a certain decision procedure, but by certain roles and functions delineated within a constitutional arrangement.

Hegel's theory of sovereignty is illustrative of this constitutionalist approach. While Hegel named a "sovereign"—a constitutional monarch—in his theory of the state, this figure did not hold ultimate, much less absolute, power. Rather, it was the entire constitutional system—the *idea* of the state, in Hegel's terminology—that was the ultimate source of authority, of sovereign power (Hegel 1991, 314–15). Monarchs, in their function as rulers entitled to make final decisions—especially in cases of interstate conflict—held power pursuant to their fulfillment of the idea that the state embodied. But it was the idea—the constitution—that was the ultimate authority.

Of course, Hegel, as an "absolute idealist," regarded the (specific) *idea* of authority as itself authoritative. But a more materialist version of constitutionalism would regard the constitution as delineating functional roles, both for sovereign, and for subsidiary, authorities. In this way, no element of a state could be regarded as having absolute power over others. The division of powers, the rule of law, and the rights and duties of citizens would thereby be preserved and protected.

The importance of a constitutional understanding of sovereignty is the fact that it recognizes other sources of *internal* power and authority than the state per se. The rise of popular sovereignty in the eighteenth and nineteenth centuries is one example of this. The development of federalism, with its recognition of substate autonomy based on geographical units, is another. The development of "functional" sovereignty, oriented to the maintenance of certain values such as ecological sustainability, is yet a further example of such sovereignty. All involve the division of power within states, in accordance with various criteria, while not seeking to challenge or undermine state sovereignty, when restricted to its legitimate role (primarily as a guardian of external independence and territorial integrity).

Despite the dawning awareness of such a change in the conception of sovereignty, much recent commentary on the shortcomings of the sovereignty doctrine continues to be based on an outmoded understanding of the concept, drawn from early modern, and especially contractarian, thinkers. Before examining some recent revaluations of the sovereignty doctrine, it is important to consider these contemporary critiques of sovereignty, since they continue to be influential, especially among those who are perhaps unaware of the different interpretations of the sovereignty doctrine just mentioned. Two criticisms of sovereignty are especially significant here—the notion that the idea is philosophically incoherent, and the claim that it violates certain cosmopolitan aspects of political rights, especially as articulated in the post-Cold War world.

The Philosophical Critique of Sovereignty

An excellent expression of the philosophical critique, which views sovereignty as an incoherent and unnecessary doctrine, can be found in Christopher Morris's *Essay on the Modern State* (2000). It is clear that Morris thinks the sovereignty concept stands or falls with the "classical"—that is, the absolutist—version of the idea. His very definition of the term indicates this: "Sovereignty is the highest, final, and supreme political and legal authority and power within the territorially defined domain of a system of direct rule. This is the core of the notion of sovereignty" (Morris 2000, 178). He adds that the sovereignty concept necessarily has the characteristics of "absoluteness, inalienability, and indivisibility." Though Morris acknowledges that the classical concept is no longer the only one, he still regards it as conceptually important for the idea in general:

> Classical sovereignty no longer is as popular as it once was. It is now widely thought that sovereignty can and should be limited . . . . So we dismiss the classical view. But we shall see that Hobbes has some arguments for unlimited sovereignty that have been influential and that are quite instructive, even though they may be wrong.
>
> *(Morris 2000, 179)*

Morris claims that if the classical view of sovereignty fails, the more limited one will as well. There are three reasons why Morris thinks the classical sovereignty doctrine is untenable: [1] there is no reason to think that authority need be final and/or ultimate; [2] no viable conception of moral norms can allow a sovereign authority to overrule moral judgments just because it is sovereign; and [3] the state is an unsuitable vehicle for sovereignty claims because of its coercive nature. Once we have considered these reasons, we will examine Morris's claim that they apply by extension to a more limited concept as well. Though Morris's critique of *absolute* sovereignty is a good one, its wider applicability—to constitutional, and to external—sovereignty is more doubtful.

The first problem that Morris identifies in the justification of (absolute) sovereignty concerns whether there can or needs to be a final or ultimate authority (or power) in political life. Morris refers to a "regress argument" in Hobbes as the standard justification for this: if there is more than one source of power, one must be more authoritative than another. If there are multiple authorities, we can put them in rank order, eventually arriving at one that is the most authoritative;

this is the sovereign. The assumption that Morris challenges is whether power is "transitive" in this way—whether one dimension of power compares to another and whether the possession of one type means a comparable degree of power in another dimension:

> Power is usually relative to some particular dimension and to some given domain; power relations are multidimensional . . . . Something might be greater in power than something else in one regard but be lesser in another . . . . For instance, the power of the three branches of the American government are multidimensional in this manner.
>
> *(Morris 2000, 184)*

The conclusion is that, given the nature of real power relations—not always transitive, not always comparable—there is no reason to think that a Hobbesian sovereign is required for *any* type of authority to hold in a society: "it would no longer be true that for any finite number of powers, there must be at least one, greater than which there is none" (Morris 2000, 187).

Morris's second criticism concerns the amorality of political authority. Does political power also confer moral authority? There is no reason to think that it does:

> States are morally limited, and this is one of the main reasons for wanting them to be legally limited . . . . For [Hobbes and Rousseau]. . . the will of the sovereign *determines* what is morally right and wrong. Virtually no contemporary moral theory could endorse this claim.
>
> *(Morris 2000, 195)*

The reason, of course, is that moral norms are not justified because they are enforced by powerful individuals or institutions, but by a process of moral reasoning, of which states are generally not capable.

Even if states were not regarded as conferring moral authority, there is a concern that the exercise of sovereignty would violate moral norms. Sovereignty has an instrumental value for the procurement of public goods that is unavailable to persons on their own. As Morris admits, this seems to be the reason most favored by political theorists for granting a sovereign power authority over persons (Morris 2000, 209). But this need for public goods does not apply equally to all individuals. Some will be capable of providing for themselves; and for them, sovereignty is therefore not justified instrumentally. But this seems little more than a hypothetical worry: the fact that a few individuals can provide for their needs privately does not mean that a state would not be justified in acting to ensure that most have similar access to these goods.

A third reason that Morris gives for why sovereignty is questionable is that states cannot provide persons with adequate reasons for obedience independent of other reasons also obtaining. Morris refers to Leibniz's criticism of Hobbes: individuals could justifiably trust the state to make decisions only if the state had the knowledge, authority, and power of a god (Morris 2000, 209–10). Not only is there no reason to trust a state in this way; there are good reasons *not* to do so. Above all, there is the reason that they are "large and bureaucratic organization[s]." Morris approvingly quotes Leslie Green's conclusion from this fact: "The authority of the state is problematic in large part because it is the state's" (Green 1988, 59, quoted in Morris 2000, 211).

This does not seem to be a particularly strong argument, since it would have to be shown that sovereign states *never* serve people in generally useful ways. But Morris also offers a different reason for this criticism: since it is *possible* to conceive of social life without states, it may be *reasonable* to do so. As Morris writes,

> [There] seems to be a fair amount of agreement that *some* social order can be secured independently of centralized political power or modern states . . . . It is important for social order, and for many other social goods, that collective decisions be made and that conflicts be resolved in an efficient and predictable way. This fact does not require state sovereignty, however, and it may not even, under some conditions, require states.
>
> *(Morris 2000, 217–18)*

However, as Morris admits, societies without institutions holding political power would be "fragile in certain ways and subject to sudden and unpredictable changes." The *possibility* of foregoing sovereign authority does not seem a good reason to actually do so.

Regarding sovereignty as constitutional, however, may escape some of these criticisms. Or will it? Morris's view is that limited, constitutional sovereignty is not all that different in crucial respects from absolutist sovereignty. For instance, if sovereignty pertains to the state, and the state is not be trusted, limiting its power will not fundamentally change this concern:

> It is hard . . . to see how the limited—but nevertheless extensive—sovereignty claimed by states could be justified, even if we suppose them to be reasonably just and minimally efficient . . . . It is difficult to imagine that this authority could be justified with regard to *all persons* in the state's jurisdiction.
>
> *(Morris 2000, 209, italics added)*

Again, the worry is that, since authority requires coercion at some point, such coercion will be unjustified in at least some instances in which persons do not need what the state seeks to provide. But why this invalidates the authority of a state or sovereign *in general* remains unclear. It is an assumption of most political theorists that collective actions require coercion of individuals in some instances, even when those individuals assent to the ultimate good that is the purpose of the actions (as well as some cases in which they do not). The fact that sovereignty legitimates such coercion does not mean that it is not susceptible to limitation and evaluation.

Finally, Morris seems to think that, even if sovereignty were given a constitutionalist interpretation, it would not solve any important problems—including the avoidance of constitutional crises and challenges of various kinds to state authority (Morris 2000, 220–21). Invoking the sovereignty doctrine to deal with such crises of authority solves nothing, since it is the doctrine itself that is in question. Furthermore, when a constitutionally mandated division of powers is instituted, invoking sovereignty claims internally does not settle which parts of a political community can legitimately claim which powers or what authority. It is the role of the constitution, not the decision of a sovereign actor or agent, to settle that.

There are two ways to understand this point about internal sovereignty, though neither implies that sovereignty per se is no longer an applicable doctrine. On the one hand, as Hegel put it, the constitution *is* the ultimate authority, the sovereign, not an office or a person within the political community. On the other hand, sovereignty may be viewed as "latent," in the sense that it need not be invoked in the determination of which agencies have what powers (Grimm 2015, 71).

Beyond this point about internal sovereignty, there is also the fact that sovereignty claims do not in every instance apply only to states. They may also be claims made by *other* political communities—very likely substate communities of one sort or another asserting their own sovereignty. So it is very important to get clear about *which* political authorities have sovereignty over *what*. This is part of what a constitution is designed to do, of course. But constitutions themselves need justification on the basis of particular interpretations of political doctrines, including sovereignty.

The general philosophical critique of sovereignty relies, therefore, on a conflation between absolutist and nonabsolutist sovereignty—there are important differences between them. Second, it also conflates state sovereignty with political sovereignty more generally. They are not the same, and differentiating between them functionally is an important part of a contemporary doctrine of sovereignty. In more recent work, Morris has fallen back on the claim that the conditions for achieving real sovereignty have simply become too hard. In competition with other sources of power, and by attempting to manage a large, complex society, aspirants to sovereignty will inevitably fail to secure ultimate authority (Morris 2021, 28). But everything in politics ultimately requires negotiation between competing sources of power: why should sovereign authority, suitably circumscribed, be an exception?

In any case, Morris seems to miss the nature of what in earlier work was called the "paradox of sovereignty"—that is, that a given political community lacks sovereignty only if there exists another political community that is sovereign instead (Dahbour 2006, 113). Because authority, and in particular ultimate authority, seems to be a constitutive feature of collective and political action in general, some political community will always effectively have sovereignty—the important question is whether they have it legitimately. The very existence of communities, in other words, implies the correlative doctrine that such communities—that is, their governing institutions—are sovereign, at least over some domain of social action. To deny that a particular political community is sovereign is simply to maintain that there exists a different one that is effectively or legitimately sovereign instead.

When external sovereignty is taken into account, a different sort of consideration is involved in understanding the limitations of sovereignty—without, however, meaning that sovereignty as such no longer exists, or is ineffective or irrelevant. It is actually an *entailment* of the sovereignty doctrine that states have obligations to other communities—specifically, to respect *their* sovereignty. F. H. Hinsley has made this feature of sovereignty—the complementary but different aspects of internal and external sovereignty—a central feature of his characterization of the doctrine. In his account, part of the purpose of the doctrine is the *limitation* of a political community's authority to their own internal affairs. Furthermore, the idea that there could be a sovereign authority "above" or beyond that of particular political communities is also refuted by the doctrine:

> In the context of the internal structure of a political society, the concept of sovereignty has involved the belief that there is an absolute political power within the community. Applied to the problems which arise in the relations between political communities, its function has been to express the antithesis of this argument—the principle that internationally, over and above the collection of communities, no supreme authority exists.
>
> *(Hinsley 1986, 158)*

This interpretation of sovereignty has implications for "cosmopolitan" views that regard the doctrine with suspicion. It is this cosmopolitan critique, even more than the supposed philosophical quandaries that Morris raises, that has often been seen to call the idea into question. Yet, the basis of this critique seems to misunderstand the very purpose of the idea of sovereignty itself—that is, the limitation of power beyond the scope of legitimate political communities (including, in some cases, substate communities).

The Cosmopolitan Critique of Sovereignty

The cosmopolitan argument against sovereignty makes three major points: [1] today, social and environmental problems are often global in scale; [2] global political institutions should be developed to deal with these problems; and [3] attempts to renovate the sovereignty doctrine will fail as a substitute for such global institutions. These points will be addressed in turn.

The idea that contemporary environmental problems (e.g., species extinction, ecological degradation, climate change), as well as the social problems that accompany them (e.g., increasing impoverishment, political instability, population displacement), are global is a commonplace. What does this mean, and what implications does it have for existing state institutions? Two points in this regard are paramount. First, there are "flows" of "political, economic, and ecological activity" that cannot be contained within state boundaries or regional locations (Kuehls 1996, ix). Conditions are produced by nonstate agencies (e.g., large corporations) or nonhuman processes (e.g., environmental changes) that have large-scale regional and even worldwide effects. Political institutions rooted in localities must scramble to ameliorate, or even monitor, such conditions. Second, the objects of such political institutions concern entities that transcend, or cross-cut, the territorial boundaries that demarcate those institutions. For example, a goal of environmental protection is the preservation of ecosystems; yet, the boundaries of ecosystems do not generally correspond to state borders. It follows that any attempt to address environmental degradation only within existing states will fail (de-Shalit 2006, 84). Similar problems exist with regard to economic development, where economic exchanges often operate across regions, rather than within states, or to demographic change, which often involves population movements across state boundaries.

Two consequences are often drawn from these observations. On the one hand, the ability of sovereign states to deal with environmental and other problems of this kind is weakened by the very nature of such problems, which undermine and erode the power of such states (Litfin 1998, 3). This "erosion-of-sovereignty" idea therefore tends to validate the notion that problems that overstep state boundaries need supra-state agencies to deal with them. Furthermore, these agencies must be animated, not by local interests or concerns, but by cosmopolitan values that accord equal weight to all persons globally. As David Held, one of the major figures articulating this view, puts it succinctly, "Only a cosmopolitan political *outlook* can ultimately accommodate itself to the political challenges of a more global era" (Held 2003, 523).

Cosmopolitan values in this context means *individuality* (separate persons, not peoples, countries, or nations, are the basic units of ethical concern), *universality* (*all* persons, irrespective of class, ethnicity, or religion, are equivalent objects of concern), and *generality* (persons are to be of concern to *everyone*, not just to their relations or compatriots) (Pogge 1992, 48–49). Together, they constitute an imperative for *moral* agents to act in ways that ensure all persons have respect and dignity.

The paramount question concerns what *political* institutions are required for this moral imperative to be acted upon. Transnational institutions, of course, already exist; the question is whether they are suitable for the realization of cosmopolitan ends, and, if not, whether they can be made to serve through suitable reforms or modifications. With the creation, most importantly, of the UN (but also of the World Court and other international judicial bodies, as well as of quasi-global economic institutions such as the International Monetary Fund and World Bank), "no member state is as sovereign as states had been in the Westphalian order. There is now a public power above them" (Grimm 2012, 130).

Clearly, if certain problems—economic, environmental, demographic, security—occur on a global scale, then transnational institutions have the *potential* to deal with them at that scale. In fact, there is some evidence of this potential, for two reasons. For one thing, while states still play important roles in provisioning for such goods as wealth, liberty, security, and sustainability, many are not in a position to do so by themselves—they must act in concert with larger international bodies (Peters, Klabbers, and Ulfstein 2009, 198). For another, unlike states, which operate *territorially*, transnational institutions such as the UN operate *functionally*, and are therefore potentially better suited to dealing with situations of scarce resources or coordination problems (Grimm 2015, 116).

Given such potentials, and the problems that states seem increasingly unable to meet, the idea of a "post-Westphalian" community that transcends state boundaries and sovereign political communities seems attractive. Such a community needs to be created, on this view, to overcome "the enduring political fragmentation in the world [which] is at variance with the systemic integration of a multicultural world society and is blocking progress towards civilizing relations . . . between states" (Habermas 2012, 7). And, as various commentators have noted, since elements of such a community are being created, sovereign territorial states no longer possess a monopoly of political power, as they might once have done.

But it is also the case that global governance institutions, while they may seem to be more appropriate for the solution of transnational problems, have a "legitimacy problem"—they are not legitimated in the same way or to the same degree as territorial states. Such institutions have sometimes been created by treaties between states; but even in these cases, the resulting bodies subsequently act independently of any procedure that would legitimate their policies, acts, or initiatives. Often such bodies are established by states acting in a partisan capacity to ensure their continuing dominance in an arena of international conflict (e.g., NATO or, more recently, the International Criminal Tribunal for the former Yugoslavia). In either event, this means that such institutions have no greater legitimacy than the states that established them.

There are two aspects to this problem—one is the exclusionary character of allegedly global institutions, the other is the internal functioning of such institutions. In the first case, an institution acts to exclude concerned parties from participation (e.g., the exclusion of Cuba from the Organization of American

States); in the second, the institutions claim universality but the structure of decision-making favors an elite (e.g., the permanent Security Council members in the UN) (Zolo 1997, 101–02).

Whether international institutions operate as proxies for territorial states or as nominally independent, but actually subservient agencies, there is no reason to think that they will necessarily act to realize a cosmopolitan agenda. Yet, as Andrew Linklater has noted, this is a precondition for achieving elements of such a program: "The post-Westphalian era will begin [only] when societies act as cosmopolitan citizens who aspire to make progress together towards the ethical ideal of a universal . . . community" (Linklater 1998, 211).

Perhaps popular movements could pressure international bodies to democratize, and thereby provide the preconditions for a truly cosmopolitan community. But there are two problems with this approach. First, what is the mechanism by which—whatever the popular demands for it—democratic initiatives will be institutionalized in global governance? Since such mechanisms do not presently exist, the standard answer is to hypothesize their existence as local institutions expanded to the global scale. For instance, Daniele Archibugi suggests the possibility of global political parties existing within the framework of a world parliamentary structure (Archibugi 1998, 198–228). But this raises the issue of whether, aside from the general problems with political representation as an adequate form for popular initiatives, one could view political parties as anything other than mechanisms for elites to consolidate their rule (Hawthorn 2003, 19–20). And if so, how could "global" parties not simply represent "global" elites, much as local political parties often do for local elites? (It should also be clear that nongovernmental organizations, however useful for highlighting certain issues, represent no one but their own members, officers, and donors, and are ill-suited to serve any kind of institutionalized democratic role.)

If this problem can somehow be addressed on the local level—perhaps by the influence of political movements or neighborhood activism—will this be equally possible on the global level? Here, the difficult problem of the relation of political democracy to geographical scale should be considered. The standard view is that, as the scale of the polity increases, the potential for popular rule decreases—though the ability of a state to protect its autonomy from others increases. There is therefore an inverse relation between the capacity for autonomous action by states, and their democratic accountability—the larger the state, the more autonomous, yet the less democratic. But it should be noted that the range usually considered is that from city-states to territorial states ("nation-states")—"global governance institutions" are generally not considered at all (Dahl and Tufte 1973, 135).

Nevertheless, we can extend the argument. What reason is there to think that, as governance is extended to the regional, and then global, scales, democracy will become any more possible than it already is on the territorial scale—which is to say, not very much? In fact, it seems probable, as Nadia Urbinati has emphasized, that the inverse relation between geographical scale and political participation will be ramified:

> The extension of democracy beyond state borders . . . allows for more participation, but can also give rise to a proliferation of powers that de facto *decrease* the chance of an effective control and coordination, and finally participation itself.
>
> *(Urbinati 2003, 81–82)*

While in theory, democratizing international bodies such as the UN could expand the participating parties, in practice, it can allow for greater manipulation of weaker participants by stronger ones. Alternatively, expanding participation in such bodies can render them less effective, while the real work of the institutions goes on informally or secretly.

These problems are now acknowledged even among advocates of "cosmopolitan democracy" (Gould 2004, 173). They have played an important role in the "sovereignty debate." If global institutions cannot be democratized, then sovereignty may remain a sign of democratic initiatives *against* international bodies that are unaccountable and unreformable. As Martti Koskenniemi has written, "Today, [sovereignty] stands as an obscure representative against disillusionment with global power and expert rule" (Koskenniemi 2010, 242).

But even if international institutions cannot easily be democratized, perhaps they can be *constitutionalized* so as to build in a more determinate set of goals and functions that could have a clearly cosmopolitan purpose. The aspiration to constitutionalization has arisen among jurists and others who see the weakening of state sovereignty without a corresponding increase in the legitimacy of international bodies. The "quest for constitutionalization" is now often regarded as the best possible antidote to the "erosion of sovereignty due to globalization" (Peters, Klabbers, and Ulfstein 2009, 182). Yet, as even a proponent of constitutionalization, Seyla Benhabib, admits, it does not necessarily solve the legitimacy problem that plagues global institutions. This is because changes in global legal statutes are being made, not by peoples but by political elites acting either on behalf of powerful states, or through transnational organizations. As Benhabib states,

> We should be concerned that the rush to embrace global constitutionalism . . . is leaving the question of democratic legitimacy unanswered. Post-democracy . . . attenuated through the systems of anonymous governance initiated by global specialists, are becoming increasingly attractive.
>
> *(Benhabib 2011, 121)*

Advocates of constitutionalization argue for limitations on the hegemonic powers who dominate international institutions through the establishment of constitutional guarantees that non-hegemonic states will have equal inputs into decision-making, whether in the UN Security Council, the International Monetary Fund, or other bodies. The normative argument for this is based on the "cardinal principle" that all sovereign states possess "equality in the making of public international law" by virtue of their recognition by other sovereign states (Cohen 2012, 283).

This is not, however, a sufficient reason for thinking that the constitutionalization of global governance institutions will develop parallel to the constitutionalization of states since this "tends to neglect the significant differences between a full-fledged constitution and the present state of international law" (Grimm 2015, 107). These differences are of two kinds—the formal differences between such international documents as the UN Charter and most state constitutions, and the substantive differences between popular movements that have historically engaged in constitution-making and nongovernmental organizations that seek to democratize global governance today.

In the first case, Jean Cohen has enumerated the problems with using the UN Charter, the document most relevant to a global process of constitutionalization, as the basis for such a process. Problems include: the absence of a court system with compulsory jurisdiction to enforce compliance or settle disputes on the global level, separation of powers or internal mechanisms for checking abuses of power, and effective protections for basic rights. Her conclusion is that "the Charter as a *constitutional* treaty is deeply flawed, and seriously deficient from a qualitative constitutionalist perspective" (Cohen 2012, 290–91).

In the second case, the major historical instances of constitution-making involved the contestation for control of territorial states between putatively sovereign rulers and popular movements struggling to institute a constitutional system of limited powers within the state. But global governance institutions today do not control governmental bodies with definite territorial jurisdictions, so much as play certain nonessential functions between or over existing states. Such international institutions were largely established by hegemonic states, which are not likely to allow the transformation of these bodies in ways that would challenge their hegemony (Zolo 1997, 8). A more likely outcome resulting from attempts by such bodies to increase their power would be their weakening or even collapse when member states withdraw their support, as happened to the League of Nations in the mid-twentieth century.

Faced with such eventualities, some cosmopolitans have turned to regional entities, especially the European Union (EU), as a venue for realizing their ambitions. For instance, Jürgen Habermas argues that the EU can serve the purpose of a "transnationalization of popular sovereignty" in which the "EU citizenry as a whole shares the constitution-building power with a limited number of 'constituting states'" (Habermas 2012, 11; see also Benhabib 2011, 116).

There are two problems, however, with viewing regionalization as the proper context for constitution-making. First, there is no discernible citizenry that exists on a regional or continental scale such as that of Europe, and second, if a political community is somehow created at that scale, there are no good means for avoiding the loss of popular participation, and the corresponding monopolization of political power by elites.

In regard to the first problem, Habermas claims that there is at least no *conceptual* barrier to conceiving of a pan-continental citizenry, since "an increase in territorial scale alone—ie a *merely numerical* enlargement of the basic population of participants—changes the complexity, but not necessarily the quality, of the process of opinion- and will-formation" (Habermas 2012, 19). But this is to mistake a quantitative for a qualitative change; geographers have frequently noted that changes in scale have significant effects on political organization, among other things (Smith 1992; Sheppard and McMaster 2004). At the very least, organizing regional or continental populations for political action is much more difficult. What has happened, as William Wallace noted early on in the EU's consolidation, is a mobilization and consolidation of political *elites*, who have the wherewithal to "flourish with multiple identities, moving from state to state" (Wallace 1999, 98–99).

In terms of the second problem, it may be the case that cosmopolitan attempts to administer new legal norms independently of popular institutions of self-rule should be understood less in terms of the "onward march of great [cosmopolitan] principles," and more in terms of the "constitution of legal elites" that, for instance, "have populated the EU" since its beginnings (Koskenniemi 2010, 229). In other words, cosmopolitanism is itself a particular form of rule, with its own institutions, as well as norms. Even where it is not the case that hegemonic states largely control global institutions, these institutions operate according to the norms of legal and economic expertise, which provide the technical knowledge required to (attempt to) govern the planet as a whole. Yet, it is precisely these functional, rather than territorial, "powers" that themselves will require even greater degrees of control in order to institute effective solutions to global problems. The result, as Koskenniemi puts it, is that "these new forms of [global] power . . . *are the new sovereignty*; they enthrone managers where there used to be kings and emperors" (Koskenniemi 2010, 240; italics added).

The fact that global governing bodies are attempting to exercise power through management by elites means that the presumed decline of sovereignty has not occurred as expected. Furthermore, despite cosmopolitan hopes that *popular* sovereignty can be disconnected from particular territorial or geographical scales, it is hard to imagine this actually occurring. More likely is the continuation of conflicts between local and popular claims to sovereignty rights and attempts by hegemonic states and/or global institutions to consolidate power by their governing elites.

In the end, the ideas of global governance or cosmopolitan law are themselves concepts of political power, but ones even less legitimate than the classical notion of state sovereignty (Koskenniemi 2004, 500). Of course, this realization does not change the fact that the state sovereignty doctrine has been transformed in various ways that render the classical version of it problematic. But the idea that there has been a "death of sovereignty" is clearly mistaken; what has occurred is simply the "disappearance of the conditions for a *very specific concept* of sovereignty" (Grimm 2015, 102, italics added).

The New Sovereignty Doctrine and Its Limitations

This change in the nature of sovereignty will be a particular concern of the next chapter, in which contemporary understandings of sovereignty will be connected to the concept of self-determination, and its transformation by the colonial liberation movements of the mid-twentieth century. It is worth noting, however, that, irrespective of this change in the meanings of sovereignty and self-determination, there has been a recent renovation of the sovereignty doctrine by the political theorists Anna Stilz and Philip Pettit (Stilz 2019; Pettit 2023). This work is in many respects compatible with the present study, though it also differs in some ways.

Stilz asks the question: Is there something that (still) justifies the political rights of sovereign territorial states, in light of the criticisms of them by global ethicists in particular? Her answer is that such states are justified by "occupancy rights" that persons have to (collectively) control the territories in which they live. This occupancy right relies on a "political autonomy account" of territoriality that sees jurisdictional authority over a territory as justified by the "self-directed agency" of a group of persons (and derived from individual person's similar possession of such agency) (Stilz 2019, 105). This view has some similarities to the "purely political" account of territorial rights discussed in Chapter 4 (and held by Margaret Moore, for instance).

But there are differences. In particular, Stilz insists that collective autonomy is a right that is premised on a similar, but supposedly more foundational, right of individual autonomy (Stilz 2019, 151–52). It is foundational because Stilz fails to find any compelling value to collective autonomy (and a corresponding right to self-determination), apart from the preferences or desires of groups for association. This is, in part, because she rejects the primacy of such association, understood as "peoplehood." That is to say that peoples are created only by conscious acts of (political) cooperation, on her account, not by any pre-existing manifestations of peoples in relation to territories, lands, or other pre-political relations (Stilz 2019, 124).

Clearly, Stilz's view differs from the account given in Chapter 4, in which peoples and territories are co-defined. Her view is helpful in providing a definite rejection of nationalist accounts of territoriality as based on self-defined (imagined) affinities with particular lands. But she goes to the opposite extreme, bringing territoriality back into close proximity to (though not identity with) proprietary theories based on private-property ownership relations (Stilz 2019, 33–34). The problem, as I argued in the last chapter, is that, in the absence of a materialist account of the relation of peoples to lands or territories necessary for livelihood and subsistence based on particular ecosystemic goods, it is impossible to establish any definite relation between particular peoples and particular places. Yet, such relations are an urgent requirement for facing the current environmental crisis that, as Kolers, Moore, and others have recently acknowledged, is a necessity.

The other somewhat puzzling aspect of Stilz's theory of sovereignty is that it concerns territoriality much more than sovereignty per se. The reason for this is not hard to find: there just is an assumption that sovereignty must be *state* sovereignty, and that its justification follows straightforwardly from the value of territoriality that she is more concerned to justify. Stilz does acknowledge criteria of self-determination rights that might lead to changes in state authority or boundaries. Two that she mentions are those of "persistently alienated internal minorities" and of indigenous peoples (Stilz 2019, 134, 138). But these cases are ones of legitimate claims to *new* states, not to forms of sovereignty that do not go the level of statehood but seek federalist or functional modifications to existing state authority.

Despite not being exclusively concerned with sovereignty, Philip Pettit's book does expound a theory of sovereignty that is, in some ways, compatible with Stilz's work on territoriality. It also accords with the critique of absolutist sovereignty given earlier. For Pettit, the absolutist conception of sovereignty establishes its importance as a feature of political (or state) authority, but goes too far in rejecting constitutional, or what Pettit calls "polycentric" versions of sovereign authority. There are advantages to "decentralized" sovereignty in avoiding the problems of authoritarian (e.g., Hobbesian) and majoritarian (e.g., Rousseauian) versions of sovereign authority (Pettit 2023, 155).

Furthermore, Pettit advocates a conception of sovereignty that is strikingly similar to Hegel's (though he references Kant, instead), in that the sovereign is the state constitution (or "polity," in Pettit's words) itself, rather than any particular official, agent, or person within the political structure (Pettit 2023, 156). Pettit calls the sovereign a "superordinate body" that "emerges from the interaction and cooperation of [the state's] individuals and agencies" (Pettit 2023, 160). The sovereign is, if not quite an idea (as Hegel indicated, and as would be embodied in a constitution), then it is a "realized" idea embedded in the various state organs.

Yet, in the end, the same problem recurs in Pettit as in Stilz, and can be put in the form of a question: why does political sovereignty necessarily always manifest as state sovereignty? The importance of the question is indicated by the limitations of states in solving certain intractable problems—environmental ones such as extractivism, above all. It is also raised by the manifestations of political authority that have developed in substate entities—indigenous peoples, political movements, common property institutions, and local governance bodies. It is important to remember why states do not constitute obvious, or even in many cases, possible alternatives if certain problems are to be confronted.

States are institutions that seek power in order to serve the interests of particular groups. Which groups? Those who have an interest in obtaining the widest possible access to material resources that can fuel and fund their control over ever increasing lands and peoples. This interest need not involve violence, though it often does. But it does always involve seeking greater wealth, in order to achieve increasing political power over groups and individuals who might serve, or conversely threaten, those seeking the power.

This is just to say that state power has a definite *content*—it is not a neutral mechanism for people's self-rule and protection. Other political groups, as enumerated earlier, may have other interests—for instance, in the preservation of natural environments. But it is improbable that sovereign territorial states will, of their own accord, act in this way, since it goes against the inherent interests that states have in increasing their own power—and the economic and military means to achieve that.

Of course, this is not to say that states' actions cannot be modified by the political initiatives of different interests, groups, or movements. But such modification, as this book is arguing, requires political sovereignty claims on the part of *nonstate* actors, in order to achieve other goals. It is now time, in the next chapter, to complete the case for understanding sovereignty in this way—not as the ultimate authority of state institutions, but of peoples embedded in, or otherwise connected to, ecosystems.

References

Archibugi, Daniele. 1998. "Principles of Cosmopolitan Democracy." In *Re-Imagining Political Community: Studies in Cosmopolitan Democracy*, edited by Daniele Archibugi, David Held, and Martin Köhler. Stanford, CA: Stanford University Press.

Arendt, Hannah. 1963. *On Revolution*. New York: Viking Press.

Benhabib, Seyla. 2009. "Twilight of Sovereignty or the Emergence of Cosmopolitan Norms?: Rethinking Citizenship in Volatile Times." In *Democracy, States, and the Struggle for Global Justice*, edited by Heather Gautney, Omar Dahbour, Ashley Dawson, and Neil Smith. New York: Routledge.

Benhabib, Seyla. 2011. *Dignity in Adversity: Human Rights in Troubled Times*. Cambridge: Polity Press.

Bodin, Jean. 1992. *On Sovereignty: Four Chapters From "The Six Books of the Commonwealth,"* translated by Julian H. Franklin. Cambridge: Cambridge University Press.

Cohen, Jean. 2012. *Globalization and Sovereignty: Rethinking Legality, Legitimacy, and Constitutionalism.* Cambridge: Cambridge University Press.

Dahbour, Omar. 2006. "Advocating Sovereignty in an Age of Globalization." *Journal of Social Philosophy* 37 (1): 108–26.

Dahl, Robert, and Edward Tufte. 1973. *Size and Democracy.* Stanford, CA: Stanford University Press.

De-Shalit, Avner. 2006. "Nationalism." In *Political Theory and the Ecological Challenge*, edited by Andrew Dobson and Robyn Eckersley. Cambridge: Cambridge University Press.

Elshtain, Jean Bethke. 2008. *Sovereignty: God, State, and Self.* New York: Basic Books.

Gould, Carol C. 2004. *Globalizing Democracy and Human Rights.* Cambridge: Cambridge University Press.

Green, Leslie. 1988. *The Authority of the State.* Oxford: Clarendon Press.

Grimm, Dieter. 2012. "Types of Constitutions." In *The Oxford Handbook of Comparative Constitutional Law*, edited by Michel Rosenfeld and Andras Sanjo. Oxford: Oxford University Press.

Grimm, Dieter. 2015. *Sovereignty: The Origin and Future of a Political and Legal Concept*, translated by Belinda Cooper. New York: Columbia University Press.

Habermas, Jürgen. 2012. *The Crisis of the European Union: A Response*, translated by Ciaran Cronin. Cambridge: Polity Press.

Hawthorn, Geoffrey. 2003. "Running the World Through Windows." In *Debating Cosmopolitics*, edited by Daniele Archibugi. London: Verso.

Hegel, G. W. F. 1991. *Elements of the Philosophy of Right*, translated by H. B. Nisbet. Cambridge: Cambridge University Press.

Held, David. 2003. "Law of States, Law of Peoples." In *The Global Transformations Reader*, edited by David Held and Anthony McGrew, 2nd ed. Cambridge: Polity Press.

Hinsley, F. H. 1986. *Sovereignty.* 2nd ed. Cambridge: Cambridge University Press.

Hirst, Paul. 1987. "Carl Schmitt's Decisionism." *Telos* 72: 15–26.

Hobbes, Thomas. 1958. *Leviathan, Parts 1 and 2.* Indianapolis, IN: Bobbs-Merrill.

Johnson, James Turner. 2014. *Sovereignty: Moral and Historical Perspectives.* Washington, DC: Georgetown University Press.

Koskenniemi, Martti. 2004. "International Law as Political Theology: How to Read *Nomos der Erde*?" *Constellations* 11: 4.

Koskenniemi, Martti. 2010. "Conclusion: Varieties of Sovereignty—Powers of a Paradox." In *Sovereignty in Fragments: The Past, Present and Future of a Contested Concept*, edited by Hent Kalmo and Quentin Skinner. Cambridge: Cambridge University Press.

Kuehls, Thom. 1996. *Beyond Sovereign Territory: The Space of Ecopolitics.* Minneapolis: University of Minnesota Press.

Linklater, Andrew. 1998. *The Transformation of Political Community: Ethical Foundations of the Post-Westphalian Era.* Columbia: University of South Carolina Press.

Litfin, Karen. 1998. "The Greening of Sovereignty: An Introduction." In *The Greening of Sovereignty in World Politics*, edited by Karen Litfin. Cambridge, MA: Massachusetts Institute of Technology Press.

McCormick, John P. 1998. "The Dilemma of Dictatorship: Carl Schmitt and Constitutional Emergency Powers." In *Law as Politics: Carl Schmitt's Critique of Liberalism*, edited by David Dyzenhaus. Durham, NC: Duke University Press.

Morris, Christopher W. 2000. *An Essay on the Modern State.* Cambridge: Cambridge University Press.

Morris, Christopher W. 2021. "Sovereignty, Some Skeptical Thoughts." In *Sovereignty as Value*, edited by Andre Santos Campos and Susana Cadilha. Lanham, MD: Rowman & Littlefield Publishers.
Peters, Anne, Jan Klabbers, and Geir Ulfstein. 2009. *The Constitutionalization of International Law.* Oxford: Oxford University Press.
Pettit, Philip. 2023. *The State.* Princeton, NJ: Princeton University Press.
Pogge, Thomas. 1992. "Cosmopolitanism and Sovereignty." *Ethics* 103 (1).
Rousseau, Jean-Jacques. 1978. *On the Social Contract*, translated by Judith R. Masters. New York: Saint Martin's Press.
Scheuerman, William E. 1997. "Revolutions and Constitutions: Hannah Arendt's Challenge to Carl Schmitt." *Canadian Journal of Law and Jurisprudence* 10 (1): 141–62.
Scheuerman, William E. 1999. *Carl Schmitt: The End of Law.* Lanham, MD: Rowman & Littlefield.
Schmitt, Carl. 1985. *Political Theology: Four Chapters on the Concept of Sovereignty*, translated by George Schwab. Chicago, IL: University of Chicago Press.
Schmitt, Carl. 2014. *Dictatorship*, translated by Michael Hoelzl and Graham Ward. Cambridge: Polity Press.
Sheppard, Eric, and Robert McMaster, eds. 2004. *Scale and Geographic Inquiry: Nature, Society, and Method.* Oxford: Blackwell.
Smith, Neil. 1992. "Geography, Difference, and the Politics of Scale." In *Postmodernism and the Social Sciences*, edited by J. Doherty, E. Graham, and M. Malek. London: Macmillan.
Stilz, Anna. 2019. *Territorial Sovereignty: A Philosophical Exploration.* Oxford: Oxford University Press.
Urbinati, Nadia. 2003. "Can Cosmopolitical Democracy Be Democratic?" In *Debating Cosmopolitics*, edited by Daniele Archibugi. London: Verso.
Wallace, William. 1999. "The Sharing of Sovereignty: The European Paradox." In *Sovereignty at the Millenium*, edited by Robert Jackson. Oxford: Blackwell.
Zolo, Danilo. 1997. *Cosmopolis: Prospects for World Government*, translated by David McKie. Cambridge: Polity Press.

6

FROM SELF-DETERMINATION TO ECOSOVEREIGNTY

State Sovereignty or Popular Sovereignty?

As was argued in the last chapter, institutions of global governance are unlikely to be able to supplant sovereign territorial states: such institutions lack legitimacy and possess doubtful capacities for dealing with environmental problems. Nevertheless, this does not mean that states instead have the right to claim exclusive, not to say absolute, sovereignty themselves. Subject to the requirements of modern constitutional sovereignty, states lack the right to exclusive political authority. Constitutional arrangements, by their very nature, apportion authority according to various criteria, none of which however can justify centralized power structures. Above all, this means that peoples, as opposed to state institutions, are entitled to at least partial authority unmediated by governmental agencies. The result of this transformation in the meaning of sovereignty is the necessity to "rethink [sovereignty] altogether outside the classical notion of the modern state" (Lipping 2010, 186; see also Murphy 1996, 111).

This chapter attempts to articulate a version of nonstate sovereignty, culminating in a statement of the principle of *ecosovereignty*. The justification of this particular idea of sovereignty will involve showing the significance of twentieth-century movements seeking "national liberation" and self-determination for any concept of political rights. But it will also require relating such political rights to the contemporary need to address extractivism and environmental destruction by those most adversely affected by them. These aspects of a contemporary concept of sovereignty may, however, seem to be little more than an argument for some form of "popular" sovereignty. The first task, therefore, is to briefly consider why the concept of state sovereignty cannot simply be replaced by one of popular sovereignty.

DOI: 10.4324/9781003568155-10

The idea of popular sovereignty—that the people is the only legitimate claimant to ultimate authority—was first forcefully introduced by Jean-Jacques Rousseau and then reformulated in the twentieth century by Hannah Arendt (Rousseau 1978; Arendt 1963). Despite these powerful statements of the rights of "the people," the idea has not served as an effective counterweight to the notion of state sovereignty. The task here is to understand why this was the case. A brief answer is that neither Rousseau nor Arendt fundamentally challenged the idea that sovereignty was essentially an attribute of a state wielding exclusive political authority. They did, however, draw different conclusions from this idea: Rousseau asserting that the people were the sole legitimate claimant to such authority, with Arendt arguing that the people expressed their will only through political organs other than states.

As Philip Pettit points out, Rousseau, similarly to Hobbes and Bodin, was an exponent of the idea that a legitimate authority should have no challenges to, limitations on, their sovereignty (Pettit 2023, 130–36). But, of course, Rousseau disagreed with the other "absolutists" about who a legitimate claimant might be. For him, the whole body of a people, united together by a voluntary and contractual agreement, was the only possible candidate for absolute authority. Once instantiated as a people, however, there was no reason to limit its authority. The legitimacy of specific decisions could only be guaranteed by the certainty that the people alone, uncoerced by any limitations or restrictions, had made them (Cohen 2010, 60–96).

One often noted problem with this assertion that the people (rather than an individual ruler) can alone claim absolute sovereignty is that the people's will (i.e., the "general will") is difficult to identify concretely and even harder to maintain institutionally. There are many, including Arendt, who thought that Rousseau never satisfactorily solved this problem (Arendt 1977, 163–65). Arendt avoids the problem by making the people a function of institutions—she generally calls them "councils"—that just are equivalent to the people (when they act collectively through these institutions). But these bodies are really only ad hoc and temporary ones, arising in certain historical moments outside the state but not in any lasting way able to constitute a permanent alternative to it. Does Arendt view them as sovereign? No—sovereignty, for her, as for the absolutists (from Bodin to Schmitt), is only a function of state power (Arendt 1963, 24 and 159–60; see also Elshtain 2008, 114–15).

Arendt in fact celebrates the *lack* of sovereignty on the part of the councils: they are forums for clarifying the intentions of peoples through speech and debate (Arendt 1963, 168). But they are not venues of power—this is retained by state structures that are alone able to act decisively to solve urgent problems. Yet, the states have no legitimacy as political agencies; only the councils have that. There is accordingly no legitimate sovereign—power held by the state should be subordinate to the popular will as expressed in councils or other institutions of peoples' power. For Arendt, the councils were where the people were free; but in order for that to be the case, "it is precisely sovereignty they must renounce" (Arendt 1977, 165).

The result of these problems with the idea of popular sovereignty was that the state remained, by default, the sole legitimate claimant of sovereignty, whatever it might mean. But what it did mean was changed by the critique of the absolutist concept of sovereignty. Its replacement by a circumscribed and constitutional authority involving division of powers was a signal accomplishment of late modern political philosophers, beginning with Hegel. On the revisionist account of Hegel, Mill, and others, constitutional sovereignty reimagines the state as a set of differentiated "powers," without according "supreme power" to any particular part. Nevertheless, the conception of popular sovereignty, however flawed in execution, did result in the delegitimation of the idea of the sovereign as an hereditary elite or dictatorial ruler. The combined effect of these innovations was to separate sovereignty as an entitlement from the ruler as an agent, and from the state as an institution (Grimm 2015, 68, 106).

One consequence of these changes is that sovereignty becomes episodic: since the state, as a continuous institution, is no longer sovereign, and the people, as a noninstitutionalized collectivity, does not act continually, sovereignty finds expression only at certain moments. These moments are the generative ones of constitutional change: "Sovereignty withdraws into the constituent power and expresses itself only in the act of constitution making. As long as the constitution is in force, it remains latent" (Grimm 2015, 71). This means that sovereignty directly entail few specific acts or policies as long as it is viewed as popular and constitutional. Today, much of the work once done in trying to define sovereignty is now being pursued by attempts to give definition and substance to the concept of *democracy* (Grimm 2015, 120).

While this is true of the idea of "internal" sovereignty, what meaning does popular government or democracy have externally? At the level of external relations between states, sovereignty must have some sort of connection—whether contradictory or complementary—to the concept of what, since the eighteenth century, has been called (political) *self-determination*. When sovereignty is "democratized," it becomes a specific entitlement of peoples in contradistinction to states. When states continue to insist on their legitimacy as sovereign powers *against* peoples' claims to self-rule, the right of self-determination becomes applicable—whether or not this is a matter internal to states. In the case of imperial or colonial states, the matter of self-determination becomes an "external" affair: in the era of "national liberation" movements, self-rule is a claim-right increasingly invoked against "foreign" powers—the colonial states, above all.

The Correlativity of Sovereignty and Self-Determination

From the perspective of early modern political theory, it may seem as if sovereignty and self-determination are oppositional norms—the former one of states, the latter one of peoples. This continued to be the standard understanding of the concepts in the nineteenth century, as self-determination was becoming a right invoked by anti-imperial movements for statehood in central, eastern, and southern Europe (Dahbour 2003, 17–57). But it remained the case only so long as sovereignty was equated with its "classical" meaning as state sovereignty. When sovereignty came to be understood as affirming the authority of peoples or constitutions, rather than monarchs or nobilities, there was no necessary opposition between the two terms. In fact, it turns out that sovereignty is required, both internally and externally, in order for self-determination to be realized. Internally, the sovereignty of, for example, the people, is required in order for it to be able to *exercise* its liberties and prerogatives, in relation to the state. Externally, sovereignty is required in order to *protect* such liberties and prerogatives, in relation to other states.

The notion of self-determination has gradually come to be regarded as denoting the core value underlying sovereignty—that of the political liberty to enact laws and policies internally, and of the political independence to acquire recognition externally (Dahbour 2013, 164–72). Today, as we have just seen, the value of sovereignty is often understood in contradistinction to that of cosmopolitan norms. Assertions of sovereignty are a way of affirming the value of democratic self-rule in the face of the inability of global institutions to allow for it. As Dieter Grimm has recently written, "Sovereignty's most important function today lies in protecting the democratic self-determination of a politically united society . . . . As long as there is no convincing model of a global democracy . . . sovereignty protects democracy" (Grimm 2015, 128).

It is also the case that sovereignty is instrumental to achieving political independence (for self-determining communities), and to recognizing the diversity of political communities regionally and globally. Each of these values stems from a particular historical articulation of the meaning of sovereignty. As Martti Koskenniemi writes,

> It might be useful to remember . . . what sovereignty meant for those who struggled against theocratic rule in early-modern Europe, stormed the Bastille at the peak of the Enlightenment, or invoked it to support decolonization in the twentieth century.
>
> *(Koskenniemi 2010, 241–42)*

First, sovereignty was a norm used to counteract the attempt to impose a unitary theological order on a Europe divided by religious confessions. States claimed the authority to determine the laws governing religious worship and belief within their borders. Ultimately, this would mean that states claimed the right to establish different legal systems, as well. The attempts, by church and empire, to regiment and enforce unitary beliefs and practices was viewed, on the basis of the sovereignty doctrine, as illegitimate. Furthermore, such attempts were no longer regarded as a justifiable cause for war. An end to armed conflict was of course the ostensible reason for the Treaty of Westphalia, which inaugurated the modern international system, with sovereignty as a cornerstone doctrine.

Second, democracy was a claim by peoples that was justified on the basis of popular sovereignty—the right of peoples to rule themselves, and the duty of governments to serve this popular rule. In this case, sovereignty was dissociated from the personal dominion of a monarchical sovereign. To the extent that monarchy was now justified, it was as a feature of a constitutional system in which personal rule served executive functions, but did not countermand popular sovereignty.

Third, the claim of colonized and subjugated peoples to political independence was an additional iteration of the sovereignty doctrine. Since colonies and other dependencies were recognized as constitutionally separate from the colonial or imperial powers, this separation was viewed as indicative of the existence of distinct peoples. Such peoples were regarded as legally separate—even though they lacked independent governing institutions. But this legal distinction between ruling peoples and subject peoples led to the view that autonomous governance was an entitlement of any and all peoples. Populations inhabiting separate territories were clearly distinct peoples in this sense, and therefore eligible for self-determination.

These different notions of sovereignty, as they developed from the eighteenth to the twentieth centuries, were legitimated by the emergent value of political self-determination. Nevertheless, there is an important distinction to be made between the two concepts. Self-determination refers to the legislative aspect of political action—the right of peoples to make laws and enact policies. Sovereignty, on the contrary, refers to the executive function of political institutions—their right to assert authority in the end, both over persons (to realize peoples' decisions) and over other peoples (to secure the conditions for continued self-determination). As Ayelet Banai and Eszter Kollar have recently put it, "Self-determination is . . . distinct in being a bottom-up notion of political power, closer in nature to a legislative (not an executive) power . . . it even carries an anti-sovereign potential. This potential is best known [where] colonial power is sovereign and is countered by a claim of the governed to self-determination" (Banai and Kollar 2021, 185). In this regard, self-determination might also assert the rights and powers of peoples against *its own state's* sovereignty, in cases of divergence between the state and the people.

Despite this conceptual, and at times political, distinction, it remains the case that self-determination requires sovereignty, while sovereignty is legitimated by self-determination. While a legacy of anticolonialism, this conceptual relation is now inherent in the institutional matrix of the contemporary era:

> The value of sovereignty . . . derives from the value of self-determination, because in a world of sovereign states the exercise of self-determination is all but impossible in the absence of sovereignty. Sovereignty grants self-determining peoples the legal, political, and institutional context in which to exercise self-determination.
>
> *(Banai and Kollar 2021, 196)*

To reiterate, when sovereignty is an attribute primarily of states, and self-determination of peoples, the two notions seem antithetical. But if sovereignty becomes a more general function of political communities of various kinds (including peoples in different meanings of the term), then it need no longer be seen this way. Sovereignty can be understood to be necessary for self-determination, inasmuch as it is an enabling condition of self-rule. At the same time, self-determination becomes, not a critique of sovereignty per se (as when the concept means the ultimate authority of state power), but its justification on the basis of some understanding of popular, community, or group self-governance. Sovereignty and self-determination, in other words, function as correlative notions, with each implying (a certain understanding of) the other.

Postcolonial Sovereignty and Self-Determination Today

How did this shift in the meanings of such foundational political norms occur? The explanation depends on understanding the effects of the twentieth-century anticolonial movements on political theory. As long as sovereignty was seen to be a justification for, and entailment of, the authority of European monarchs, it could not but be seen as opposed to attempts to institute popular rule, especially outside the European countries. But when the subject peoples of the European colonial domains began to revolt, there were consequences for *which* governing bodies were seen as legitimate. The critique of colonialism led, in other words, to a critique of state sovereignty.

Part of this change followed from a dawning awareness that the legal distinction between colonial powers and their colonies provided a basis for arguing that the colonies were nascent political communities of some sort; why not ones that were similarly self-determining? Later attempts to disguise this distinction, such as the partial incorporation of colonies into metropolitan governments (e.g., of Algeria as French *départements* or British colonies into the Commonwealth structure), failed to eliminate the distinction between colonizers and colonized.

Recognition of this fact in political theory found its most important expression in the work of Hannah Arendt, who was the first thinker to see why European colonialism had to lead to changes in the understanding of political legitimacy. Arendt saw more clearly than others that imperialism was not, essentially, a *political* system at all, since it "has its origin in the realm of business speculation" (Arendt 1948, 125). Furthermore, since colonial empires were fundamentally mechanisms for abetting economic expansion, competition, and growth, rather than any sort of legitimate and stable political representation, they inevitably engendered opposition and eventually, their own negation.

The political reaction that led to this state of affairs on the part of the colonized was, as she noted, their discovery of a need for their own political self-determination: "Wherever the [European] nation-state appeared as conqueror, it aroused national consciousness and *desire for sovereignty* among the conquered people, thereby defeating all genuine attempts at empire building" (Arendt 1948, 127). Her own hostility to the idea of sovereignty did not prevent her from seeing that its extension to all peoples without respect to culture or nationality—and as connected to their struggles for self-determination—was an important innovation in the political thought of the twentieth century.

Here is the heart of this innovation: from an understanding of sovereignty and self-determination as antithetical to one another, sovereignty comes to be thought of as finding its very justification in a principle of self-determination. The initial development of the latter concept was rooted in popular movements against monarchy in Europe. Claims to sovereignty by reigning monarchs clashed with claims to self-determination by democratic parties. Yet, with some exceptions (notably in Austria, Germany, and Italy), these conflicts were over governance in the *same* states—Britain, France, and so forth. In this period, the idea of states—whether absolutist or populist—holding overseas colonies was noncontroversial. As Daniel Philpott has written, "The Westphalian constitution as it existed between 1648 and 1960. . . was entirely consistent with the holding of colonies . . . . Colonial rule did not go unprotested . . . . But European practitioners of statecraft regarded colonial rule as legitimate" (Philpott 2001, 154).

What changed this situation was the application of the self-determination principle to the colonial world. While, with some exceptions, self-determination in Europe itself could be achieved without necessarily changing the arrangement of sovereign states—a constitutional regime in England replacing an absolutist regime there—this was not true in the colonies. To achieve self-determination meant also becoming a sovereign entity—in Mexico, or Egypt, or India. Self-determination applied to the colonial empires was not contradictory to sovereignty, but complementary: to become self-determining also meant to become sovereign. Any idea that the European colonial empires could claim sovereignty over the peoples of the colonies, while being self-governing "at home," came to seem self-contradictory. As Robert Jackson put it, "What proved fatal to the institution of imperial sovereignty . . . was the liberal political idea that there was something inherently wrong about a government that laid claim to foreign territories and populations—even if its intentions were benevolent" (Jackson 1999, 22).

One way of understanding this convergence of sovereignty with self-determination in the twentieth-century anticolonial movements is that the concepts had a common philosophical underpinning in the idea of freedom from superior authority. The sovereignty doctrine was an outcome of a long struggle against the power of the Catholic Church—via the authority of the Holy Roman Empire—to limit self-rule in individual states. Similarly, the self-determination principle mandated self-rule for the colonies, defined by the United Nations in its key anticolonial declaration from 1960 as "non-self-governing peoples" (U.N. 1960).

The result of this convergence was that sovereignty was only a legitimate claim of communities to the extent that they also recognized the self-determination of other communities, along with asserting their own. Of course, this begs the very important question of what constitutes a community that can legitimately claim self-determination as its right. But, bracketing that question for the moment, we can see that sovereignty now entails the recognition of the self-determination of (all) peoples. In some sense, it is just a matter of extending a principle that already recognizes the sovereign rule of the people to peoples everywhere, without regard to geography, race, or culture. This is not just a matter of internal self-rule, but also of external independence from domination by other states. Undoubtedly, the affirmation of external sovereignty—the denial of legitimacy to attempts by some countries to dominate others—is the most important consequence of rethinking the idea of sovereignty to make it consistent with self-determination.

This idea that hegemony, to use the current term, or colonialism, as it has manifested itself historically, is illegitimate is one way in which the sovereignty doctrine has become connected to the concept of self-determination. Sovereignty then becomes the affirmation of authority within communities *and* the criticism of those communities that dominate others. This, as Hinsley has noted, is an essential part of the concept of *external* sovereignty: "[The] idea of sovereignty [is] the idea that there is a final and absolute political authority in the political community . . . *and [that] no final and absolute authority exists elsewhere*" (Hinsley 1986, 26).

In other words, what the doctrine prohibits is any community claiming power over others. Sovereignty internally is a doctrine of the *validation* of authority; externally, it is a doctrine principally of the *limitation* of authority. It is not hard to see why this should be so: local authorities cannot be authoritative if their judgments and actions are limited or overridden by some alien but superior power. This means that, for the sovereignty doctrine to work, there needed to be a set of principles, and even of institutions dedicated to the observance of these principles, that establishes peaceful and orderly relations between communities (Hinsley 1986, 185–86). This is something like the system of international law that has developed over the last century or so. Of course, this system can be understood in different ways. For instance, it might consist of a notion of "balance of powers," where peace is maintained through arrangements between *powerful* states. But it can also be regarded as a doctrine of the "equality of (all) states"—and this is the understanding that is consistent with the postcolonial interpretation of sovereignty (Jackson 1990; Dahbour 2013, 169–72).

In the development of postcolonial sovereignty, there was, however, one crucial limitation to this doctrine of state equality that was successfully imposed by the colonial powers. This limitation was to prove essential in sabotaging attempts to redistribute some of the material proceeds and benefits from a century or more of colonial rule. In the negotiations leading up to UN Resolution 1514, "Declaration on the Granting of Independence to Colonial Countries and Peoples," eventually adopted in 1960, the most intense debate concerned a proposed clause detailing sovereignty over natural resources. The clause read as follows:

> The right of peoples to self-determination shall also include permanent sovereignty over their natural wealth and resources. In no case may a people be deprived of its own means of subsistence on the grounds of any rights that may be claimed by other states.
>
> *(Getachew 2019, 202–3)*

Because, as Adom Getachew writes, "Fears that this language legalized nationalization and expropriation of private property led many European states and the U.S. to reject this paragraph," the offending language was dropped from the final resolution. The result was that the changing meaning of self-determination, and the postcolonial sovereignty that was the result of that change, remained limited—at least in its initial, influential formulations—to the domain of formal, political independence. Attempts at that time to secure control of the material basis of that independence were unsuccessful, at least in the diplomatic arena.

There were further attempts in the 1960s and 1970s to do just that. In 1962, for instance, the UN General Assembly did pass a resolution (1803) on "Permanent Sovereignty over Natural Resources" (Eckersley 2004, 221–22). Yet, the damage had been done: self-determination as such was no longer regarded as implying "arguments for a more egalitarian global economy" (Getachew 2019, 91–92). Resource sovereignty, to the extent that it could be secured, would be the result of negotiations between newly independent states with former colonial powers that remained in possession of extractivist enterprises in their former colonial territories. Attempts to change this—such as declarations of a "New International Economic Order"—faced a daunting array of economic and military power designed to ensure a continuing neo-colonial hegemony.

The limitations of postcolonial sovereignty suggest reasons, at the beginning of the twenty-first century, to formulate a new conception of sovereignty—one that I have argued has a basis in political ecology, rather than in the discourse of economic development. The reasons for opting for such an alternative are two-fold. Postcolonial sovereignty can be understood to be equivalent to the establishment of independent states within colonial boundaries, along with policies designed to achieve a version of development based on the European model. One reason why an alternative approach is preferable is that the postcolonial/developmental project was bound to fail at achieving real political independence and sovereignty, *even if it did eventually include some claims to resource sovereignty.* Another reason is that ecological sovereignty can alone provide the possibility of realizing a more achievable version of independence, in this case through policies aiming at "alternative modernization" and environmental justice.

Why was postcolonial sovereignty unable to realize its own goals of independence from the colonial powers? In large part, this was because of the shortcomings of the economic development model. One of the goals of sovereignty in the postcolonial era was the achievement of true economic independence. But this independence was regarded not as a type of self-sufficiency, so much as one of equal participation in global market exchanges. The short-term goal was equity in terms of trade for commodities. It was thought that this would allow the accumulation of capital sufficient to fund development projects, as well as the diplomatic and military institutions necessary for real political independence.

The newly independent states were due for disappointment. Even enforcing "fair" terms of trade proved impossible. But the more fundamental problem was that the "economic nationalism" driving the resource sovereignty approach was unable to survive the long-term trade liberalization policies that favored already wealthy countries. The historical trends in the emerging global economy all pointed to increased liberalization of terms, not the protectionism and import substitution policies that so-called developing countries favored (Harris 1990, 244–68). This did not become apparent until the advent of "neoliberal" economic policies in the 1980s. But by that time, it was clear that those with comparative advantage in global markets—generally the former colonial powers, plus a small selection of newer states (e.g., some oil-producing or other mineral-rich states), had no intention of allowing more equitable terms of trade for less advantaged countries.

This postcolonial trajectory is also evident in the development discourse that came to obsess elites in the "Third World," as well as "development agencies" in the First World. Once self-determination was apparently achieved in the former colonies, sovereignty came to mean control over their economies, which, in turn, became the ability to engage in policies aiming at "modernization," development, and above all, economic growth. But what policies? In the absence of true self-sufficiency at the local level, and the inability to secure favorable terms of trade at the global level, development came increasingly to mean simply participation in commodity or financial markets. As Wolfgang Sachs writes,

> Whenever, therefore, the 'right to development' is invoked today, it is likely to be a call of national elites to be more fully admitted to the global circuit of capital and goods, and not a call for greater solidarity with the majority world beyond the hustle and bustle of the world market.
>
> *(Sachs 1999, 164)*

The Environmental Conditions for Self-Determination

The failure of postcolonial sovereignty as a means of achieving real self-determination after the exit of the colonial powers can also be understood as the result of an inability to face the changed circumstances of the "underdeveloped" world in the late twentieth century. The obsession with a productivist, growth-oriented model of development as the basis for sovereignty was founded on ignorance of the ecological foundations of human self-sufficiency. In a world where, as Sachs writes, "The wealthy twenty-five percent of humanity occupy [an ecological] footprint as large as the entire biologically productive surface of the earth . . . such a level of resource consumption cannot be imitated by the rest of the world" (Sachs 1999, 167). In other words, the growth-oriented model cannot account for the fact that the resources necessary to achieve development by imitating the historical trajectory of the colonial powers *are no longer available.*

If the postcolonial elites persist in attempting to mimic the growth model (now disguised as "sustainable development"), it can only be for the benefit of an increasingly smaller and more privileged group in the so-called developing countries. As for the rest, they are rendered invisible by the disconnection of postcolonial elites in the former colonial capitals from the hinterland of peasant and tribal peoples comprising the majority populations of the former colonies. This, in turn, suggests two aspects of an ecologically-oriented idea of sovereignty that recommend it as an alternative. First, such an idea of sovereignty would aim to employ non-growth-oriented forms of modernization and development, involving concepts of ecological sustainability and resilience. Second, it would seek to fulfill a conception of environmental justice that mandates comprehensive equity in the distribution of natural goods—including especially to rural and indigenous populations not generally included in the circuits of capital accumulation. Both require a principle of sovereignty that does not seek state power as a primary desideratum, or economic growth as a secondary one. Instead, sovereignty can be won by building on a foundation of the empowerment of substate communities.

As we saw in the paradigmatic case of food sovereignty, instituting new policies aimed at self-sufficiency for food producers and other subsistence-oriented workers requires the preservation and strengthening of environmental goods and services in order that they can be used for meeting needs locally and regionally as much as possible. That also means taking such goods and services "off-market" in the sense that they cannot be used in order to attract investment capital, or to build exchange value through production for global consumption. Such purposes, however, tend to attract state investment and regulation, in order for the state itself to benefit from increasing foreign currency holdings and in taxing investment revenues from mining, forestry, plantation agriculture, and commercial fisheries. The state stands to gain, and local subsistence producers lose, from such ventures. In order to resist such market-state initiatives, and at the same time to encourage investment in alternative agriculture and biomass conservation, peoples need to be able to claim sovereignty over their lands and waters *against* the "developmental" state (as well as against corporate land grabs).

In addition, the achievement of environmental justice, in the sense of ecological equity, requires the resolution of distribution conflicts over environmental goods in favor of subsistence usage, local needs, and ecosystem preservation. It is not enough to redistribute to the urban poor, since this only creates incentives to increase the flow of ecological refugees into unsustainable urban centers, contributing to a "planet of slums" (Davis 2006). While states can sometimes mobilize resources to deal with the "food insecurity" of city dwellers, they have been generally unwilling to support land tenure reforms that give increased access (and ownership) of lands and waters to direct producers—farmers and fishers. Such peoples require political rights over the environments which they inhabit in order to obtain, and retain, access to their goods and services. As a general principle, (agrarian) subsistence should override (urban) consumption in these cases. But it is only a political right of sovereignty for ecosystem peoples that will provide the political advantage necessary to make that happen.

Ecosovereignty therefore has both a functional meaning and a spatial one. Just as postcolonial sovereignty was based on the self-determination of non-self-governing (i.e., colonized) peoples, so ecosovereignty is based on the self-determination of peoples who are ecologically based and subsistence-oriented, yet lack effective control over the environments necessary to sustain their livelihoods. As we saw in our discussion of food sovereignty, this is a very substantial part of the world's human population, and could potentially include even more, were urban populations to be able to acquire land for subsistence purposes.

Ecosovereignty, Defined

It is a contention of this book that, without a robust conception of political sovereignty, efforts to deal with the contemporary environmental crisis, and in particular to defeat extractivism, will fail. As we have seen, it is only sovereignty rights that can justify taking the political actions necessary to halt further destruction and to reconstruct communities and environments in a more sustainable way.

In the nature of the case, sovereignty claims assert an "ultimacy" about the role of the political agents making them. While decision-making and policy-making processes will continue to involve multiple actors, those able to justify their claims to sovereignty will get the "last word." This is exactly what is needed to deal with crises, destruction, and violence on the scale of the extractivist acts currently happening.

A full statement of the ecosovereignty principle that has so far been given partially, and with different applications, follows. The most important aspect of this restatement is to detail the conditions under which the principle will apply. The principle of ecosovereignty asserts, first, that:

> *Peoples living in distinct territories (based upon ecosystems) have ultimate authority over land use, capital investment, and other matters affecting environmental conservation in order to sustainably maintain environmental goods and services (including biodiversity) in those territories (1.0).*

The principle, second, is subject to three necessary conditions:

[1] authority will be exercised by *ecosovereign authorities* such as common property institutions, local governing councils or assemblies, or other forms of governance suitable to the administration of ecosystem-based territories, with their rules and decisions taking priority over those of state and other governmental bodies that claim authority over larger domains (2.1);

[2] participation in such authorities will be determined by membership in *ecosystem peoples* who exhibit historical attachments to the territories in question, and without excluding those (perhaps distinguished by markers of ethnicity, race, religion, or other characteristics) who have similar attachments (2.2);

[3] this authority will obtain only in order to achieve the goal of *ecological sustainability*–that is, the long-term (multigenerational) maintenance of natural environments within distinct territories, the authority becoming void if other goals (such as economic growth or "sustainable development") are substituted (2.3).

Political principles such as sovereignty justify collective actions of various sorts, depending on the historical situations in which they are formulated. Earlier conceptions of sovereignty, powerful and even revolutionary in their day, no longer capture the needs of the present moment. Faced with a civilizational crisis of unparalleled dimensions, only a form of collective action that can stem the loss of vital, life-giving matter can address this crisis adequately. Political principles that do not face the crisis as their primary desideratum will only prolong and intensify the decline and fall of major societies, unable to rethink their values and institutions.

Ecosovereignty is a principle designed to meet this moment. Based on the value of self-determination for peoples that has motivated transformative political action globally since the eighteenth century, it provides a justification for political actions that seek to prevent extractivism, conserve natural goods for future generations, protect resources essential for peoples' subsistence, and build resilience into living ecosystems. But sovereignty is a necessary part of this justification because such actions must take place within the context of *distinct territories*, based upon ecosystems. Furthermore, a principle of ecosovereignty can only justify the authority of peoples in these territories if they take actions to *sustain the natural goods and services* in them.

The principle, then, has both spatial and functional dimensions. These dimensions require further comments in order to make clear the principle's scope and applicability. But first, some specification of the general range of a principle of sovereignty needs to be clarified. Once this is done, the chapter will conclude with a summary of the justification(s) given in the book up to this point for the ecosovereignty principle.

In general, principles of sovereignty pertain to three basic matters. The ecosovereignty principle is not an exception to this, though it does have a more limited reach because it justifies functional, rather than comprehensive, authority over a given territory. The relation between functional and territorial sovereignty will be discussed presently (the relation of ecosovereignty to sovereign territorial states will be discussed in Chapter 8). The three matters that are usually affected by the sovereignty principle are: [1] jurisdictional authority (as in law-making and decision-making powers), [2] membership in a community, including rights of inclusion and exclusion, and [3] control over (natural) resources (Vanderheiden 2020, 185–86). All of these dimensions of sovereignty pertain to ecosovereignty in certain ways.

First, bodies that can legitimately claim ecosovereignty (rather than states) will have jurisdictional authority over anything that is necessary for maintaining the integrity and health of ecosystems that support human and nonhuman life in the relevant territory. That means that local populations in these territories will be able to make laws or rules to protect such natural systems, and to enforce them through decisions concerning particular uses of these systems. For instance, ownership of lands or waters, along with rules concerning the use of their living inhabitants and nonliving environments, will be decided by the governing bodies that are sovereign over them—not by larger states or agencies. What these governing bodies will be, as indicated in the second condition of the ecosovereignty principle (i.e., 2.1 earlier), may vary according to local political cultures and geographical particularities. Their jurisdiction will, however, be exercised by *ecosovereign authorities* of some sort, in accordance with local customs and cultures (e.g., village commons in Europe, *panchayats* in South Asia).

This concept of ecosovereign authorities is meant to evoke common property institutions and other local governing bodies that are capable of acting to maintain sustainable environments. For reasons given earlier, state institutions are unreliable agents for such a goal. What such authorities are, and how to make them effective, first, differs by locality, and second, is a matter for constitutional design, rather than philosophical justification. But the important point here is that states have shown themselves to have other agendas that render them unsuited for this role. Of course, the relation of such authorities to states is a crucial matter, if the ecosovereignty principle is to be realized.

The second aspect of sovereignty concerns who can claim membership in a community, and who has the authority to accept or reject, include or exclude, such persons. It is certainly the case that membership in *states* has historically been a highly charged and contentious right. A number of political theories—proprietary, nationalist—have made membership a function of individual entitlements based on acquisition, or a result of ascribed characteristics denoting particular social identities. The theory of ecosovereignty rejects such views, as the discussion of such theories applied to territorial rights in Chapter 4 made clear. Instead, the theory relies on material connections based on geographical and historical occupancies of land and territory, to determine membership in an ecosystem people who can make legitimate sovereignty claims.

Those who have a connection to territories *essential for their livelihood in that place* can claim such a connection. Similarly, those who have, or have lost within living memory (i.e., three generations), a connection to territories *necessary for a distinctive (and ecologically sustainable) form of life* can also make such claims. These are the bases for the core membership of ecosystem peoples (as in 2.2). However, there are others who may, under certain circumstances, qualify. Two examples can be given easily; similar ones may occur in other situations.

First, there are cases of peoples presently living in territories who, though they do not have a deep historical connection to the place, nevertheless seek to remain and live there according to ecologically sustainable rules and customs. (An example might be white farmers in South Africa seeking to remain after the institution of democratic governance.) Such peoples have a prima facie case for consideration as members in an already-existing ecosystem people, since their lack of ascribed connections to other, more historically rooted groups, is not a relevant criterion. There are limitations in such cases, however—in particular, the "carrying capacity" of an ecosystem, and the need to maintain its integrity over time without exhausting the natural goods and services that it provides.

A second case might be peoples wishing to move to territories that have been largely (though almost never entirely) vacated by their former inhabitants. In such instances, peoples wishing to move to, and live within, such territories sustainably—for instance, farming peoples seeking land that has been largely abandoned (except perhaps by industrial farming operations) may have a prima facie case to occupy and live in such territories. (A case of this could be landless workers in Brazil seeking to occupy parts of large estates used for industrial agriculture.) In order to do this, however, they must accommodate remaining inhabitants in mutually acceptable common property arrangements for managing the land. In all such cases, the important criterion for consideration in determining the nature of an ecosystem people is their congruence with historical and sustainable patterns of land use.

Despite these sorts of cases, the definition of ecosystem peoples just given would nevertheless exclude settler populations of various kinds, whether from near (e.g., urban homesteaders) or far (e.g., foreign settlers), unless the peoples already resident there agreed to include them. At the same time, peoples inhabiting ecosystems who are themselves divided according to ascribed identities—religion, ethnicity, and so on—would need to find ways to share power and control over the territory. Such cases as Muslims and Hindus in Kashmir, or Arabs and Jews in Palestine, could find ways to federate or power-share in order to maintain the ecological sustainability of the territories that they inhabit equitably. Of course, there is no certainty that this will happen; but the legitimacy of sovereignty claims in such territories would depend on finding and instituting such political arrangements for sharing power.

The third general feature of sovereignty is the right of control over natural resources in a given territory. As discussed earlier, one of the points of contention in the anticolonial struggles was whether new postcolonial states were able to claim *permanent* and complete authority over resources. Is it justifiable, perhaps on different grounds, for ecosovereign authorities to claim this kind of control over resources? The relevant point here is not whether such control would be permanent or limited in some way. More important is the purposes for which the resources would be used. Ecosovereignty asserts that some purposes are legitimate—those contributing to the maintenance of human livelihoods and ecological conservation. But others—especially those that use resources either for out-of-area consumption or for growth-oriented development projects more locally—are illegitimate from an ecological sustainability perspective. This perspective is the primary desideratum of the ecosovereignty principle in the first place, and therefore provides a criterion for limiting use of natural goods. If these limits are respected, then resource sovereignty is accorded to the residents of a territory. But if they are not, then (as stated in 2.3) such sovereignty is forfeited.

The uniqueness of the ecosovereignty principle is that it has functional and spatial elements that differentiate it markedly from earlier iterations of sovereignty rights. In terms of the functional aspect, sovereignty as a right is "unbundled," in the sense that sovereignty in one functional domain (e.g., land use and resource conservation) does not extend to other domains (e.g., mass communications or public education). As for the spatial dimension, control over territory by ecosovereign authorities does not mean that such control extends to the "external" level—that is, in terms of relations to other sovereign territorial states. The extent of ecosovereignty rights lies in the substate domain—in relation to states themselves, and to other regions or territories within a state that may have different (local) authorities. Some further comments on each of these aspects follow.

The authority conferred by the ecosovereignty principle has one clear function: the maintenance of ecosystems in health, at scale, and long-term, and the protection of them from extractivism. This authority is *not* a type of resource sovereignty, prioritizing local usage; *however, the resources are used.* The point of the principle is that specific uses must be sustained across multiple generations (how many determined by local beliefs and customs). Other uses are proscribed and their acceptance means the principle has been violated, and that the holder of the sovereignty right no longer has legitimate authority in that domain. (What happens then as a factual matter depends on who is able to take responsibility for maintaining ecological sustainability in that territory, and whether they can assume authority over that ecosystem.)

The question of whether sovereign authorities are able to maintain control over certain functional roles, but not others, is a legitimate one. The promise of the sovereign territorial state was that one agency could hold ultimate sovereignty over all domains of the people in a territory—and that this would provide the foundations of a secure and prosperous environment for that people. As a general matter, this promise has not been fulfilled. There are, of course, some states of a certain scale—generally, quite small ones—that have been able to do something along these lines, largely because they govern fairly uniform territories encompassing singular ecosystems. But this supports the case for ecosovereignty inasmuch as it shows that sovereignty for peoples inhabiting particular ecosystems can constitute viable and sustainable communities.

Nevertheless, for most states, ones that comprise two or more ecosystems and the peoples that inhabit them, inequitable institutional arrangements favoring one region over others has often been the norm. Such arrangements have involved taxation and investment schemes that support intensive development of some regions supported by the extraction of resources from other regions. States have often been instrumental in furthering such unequal development, often in tandem with a new business elite that is profiting from such schemes.

The antidote is a devolution of authority to particular peoples in distinct regions so that they can maintain the resources of the places they inhabit for their own livelihoods. What this requires is not only a devolution of sovereignty but also a functional sovereignty that is especially concerned with the livelihood needs of peoples, while leaving other concerns to state agencies better able to handle them. This is what the ecosovereignty principle is designed to achieve.

The principle also seeks to accomplish ecological sustainability through territorial rearrangements. It should be emphasized that "rearrangement" refers not to changes in statehood, boundaries, and citizenship but to alterations in the "internal" sovereignty of states. Is this possible? A short answer would refer to the history and philosophy of federalist governance, which is one way of devolving authority and dividing governmental powers in order to accomplish goals that centralized and absolutist sovereignty would eschew (Dahbour 2013, 193–205). It is one purpose of federalism to do just this: allocate authority to units of states better able to address certain problems of resource use, development needs, and land tenure than the states as a whole could do. There is nothing inherent in the idea of sovereignty that makes it applicable only to indivisible units (such as states).

At the same time, it might be thought that, even if sovereignty can be functionally limited in this way, the "territorial integrity" of states remains a necessity, given the dangers of secession, annexation (by other states), or even civil war. Certainly, "territorial integrity" is a basic desideratum of international law, and rightly so (Walzer 1977). If there is any political agency that might offer an alternative to sovereign territorial states, it might very well be super-territorial political agencies—ones larger and more powerful than single states, rather than smaller, substate entities.

But the possibility of super-state entities as a viable alternative to states was criticized in Chapter 5: such entities lack political legitimacy, due to their dependence on unrepresentative elites. In addition, such super-state governance agencies also lack efficacy, due to their dependence on the more powerful states, which in any case have generally manipulated them for their own purposes. But what of substate entities that might pose dangers to people living in otherwise peaceful sovereign territorial states? The danger seems exaggerated since the purpose of ecosovereignty is not to *challenge* state sovereignty generally, but to operate within existing states, aiming to better ensure ecological sustainability and livelihood sufficiency for peoples in substate, but environmentally distinctive, regions. None of the movements that seek what is here called ecosovereignty—especially the most influential ones discussed in Chapters 1 and 2 (i.e., the food sovereignty and indigenous rights movements)—seek to displace or dismember states. They hope instead to achieve their ends without needing to construct wholly new state structures. It is the internal sovereignty that existing governments claim that needs to be altered, in order to allow better protection and maintenance of ecosystems and the peoples' livelihoods that depend on them.

It is, however, the case that, if there is ever a legitimate reason for "state death"— that is, for the breakup of an existing state, usually accompanied by the secession of one or more regions, it is mandated by the persistently inequitable distribution of goods and resources within that state. This is the case that has been made in previous work, as well as by others (Dahbour 2013, 51–54). Allen Buchanan, for instance, has developed the useful concept of "discriminatory redistribution" to describe situations in which states enable or perpetrate the redistribution of goods and resources from one intrastate region to another. In the absence of the possibility of redress within an existing state's political system, this may (other conditions being equal) justify secession of the region suffering discrimination (Buchanan 1991, 1997).

One rationale for the development of the ecosovereignty principle is to provide an alternative means of dealing with such situations other than secession or partition, which often have, even in justified cases, a number of undesirable consequences. The devolution of authority, in particular along the lines of place-based autonomy concerning the maintenance of natural goods and services, seems clearly preferable as a means of ensuring an equitable distribution of such goods.

The territorial consequences of ecosovereignty then are not to destabilize the "external" sovereignty of states, but to rearrange their internal sovereignty in a productive way that allows other political communities to express their needs and protect their resources for fulfillment of those needs. Obviously, extractivist practices that plunder such resources for profit and serve distant consumer markets are antithetical to such sustainability-oriented policies. The need, therefore, for a different approach to territorial rights and governance should be clear.

Ecosovereignty, Justified

It is time to give a more systematic justification for the ecosovereignty principle just defined. This justification will take a somewhat different form than the myriad considerations leading to the formulation of the principle that has been discussed in previous chapters. It consists of three interrelated claims: [1] ecosovereignty represents the most developed form of political sovereignty for the current era; [2] the importance of such a principle of sovereignty lies in its necessity for ensuring rights to territory; and [3] territorial rights themselves are an essential element in achieving environmental justice, and in facing the comprehensive environmental crisis that is manifesting itself today in many regions of the world.

As argued in this and the last chapter, sovereignty is a concept whose meaning has changed markedly in the last century. Today, it is a notion that is closely related to the idea of self-determination, and one that is used as a critical tool against attempts to create a "new colonialism" of global governance institutions that seek to override that of independent peoples and states. Yet, this use of sovereignty does not exhaust its role in political struggles—especially *within* states that are experiencing conflicts over development, environment, and resources—that is, "ecological distribution conflicts." In such cases, assertions of sovereignty may be made against the state, "from below." "Environmental justice" movements, including those discussed in the first two chapters of this book, have employed sovereignty claims to fight extractivism in their homelands and local environments.

This is a new use of sovereignty, but not one that is inconsistent with the evolution of the concept over the last century. In fact, it is the *best* understanding of how sovereignty is correlative with self-determination, and necessary in order to realize the promise of overcoming the legacy of colonialism. The reason is that ecosovereignty is a concept that can be used to struggle against *internal* colonialism within postcolonial states; it is not identical to, or compromised by, its use by those states themselves. As we have seen, "internal" sovereignty becomes "latent" within a constitutional state, where conflicts within the state over territory and resources are fought out between and among various state "powers" (e.g., executive and legislative).

But what the concept of ecosovereignty does is to reintroduce the one, irreducible element of sovereignty that is always present—its "external" dimension—*within* states, and over contested rights and goods. As Dieter Grimm writes about this element of sovereignty, "where it refers to the external dimension, the concept of sovereignty cannot simply be replaced" (Grimm 2015, 121). Here, however, the "external" relation is not between the state and other states or global agencies, but between ecosystems, self-subsistent communities, and indigenous peoples, on the one hand, and states, on the other. It is the *same* type of (sovereignty) relation, but at a different geographical scale. And it is the local ecosystems, communities, and peoples that here can legitimately claim sovereignty against the state on the basis of their own right of self-determination.

Why is this conception preferable to earlier versions of the sovereignty principle? Because it encompasses, indeed highlights, the ecological dimension of self-determination. As discussed in different ways throughout this book, the projects of, first, colonial liberation, and later, economic development, for much of the world—including impoverished regions of many so-called developed countries—have failed to deliver on their promises. In actuality, they have often been devices for further disempowering and impoverishing subject or marginalized peoples (including, of course, many indigenous communities), while simultaneously contributing to the growth of a new global elite of consumers, managers, and professionals in both the former colonial and newly independent states.

The crucial failure of earlier conceptions of self-determination is their inability to understand that the models of development, abundance, and independence that provided certain satisfactions to elite populations in the colonial period *are no longer available to others now.* There is a simple reason for this: there is not planet enough to provide the resources to bring even a small number of the billions of humans alive today to a level of material wealth that the wealthy few attained over the last hundred years or so. Even to maintain this consumer elite in the ways of life to which they have become accustomed will destroy much of the remaining physical and biological wealth of the earth over the next century.

For the marginalized and excluded among the world's peoples, self-determination must mean protecting or obtaining *direct access* to the natural goods and services that can still provide sufficiency in livelihoods. It is sovereignty over these goods that is necessary to live self-sufficiently and to remain in places where this is possible.

The most crucial aspect of sovereignty claims understood in this way is authority over the territory and land necessary to achieve sufficiency. Without this territory, peoples will be ruled by states or global institutions charged with supporting *and controlling* subject populations. In short, without sovereignty over territory necessary to achieve livelihood sufficiency, self-determination will be impossible. Yet, this connection has gone largely unremarked in much—though not all—of the revived discussion of territorial rights in the last few years.

As I argued in Chapter 4, much of the initial revival of interest in territoriality occurred absent any consideration of the material/environmental bases of peoples' attachment to, and interest in, territory. This was what in earlier work was called the "ecological blindspot" in debates about territoriality (Dahbour 2019). While this is no longer the case, philosophers continue to argue for rights to territory on other grounds, attempting later on to integrate environmental considerations with the earlier "foundations" of territorial rights. That the ecological context of human life is *itself* foundational for the relation to territory seems still to be an insight that is hard to incorporate with other philosophical commitments.

Anna Stilz's attempt to develop what she refers to as a "foundational" account of territoriality is an example of this. She rightly emphasizes that proprietary views of territory focus excessively on the ways in which territorial rights are similar to property rights. Yet, as she indicates, it is similar inasmuch as both require a foundational—that is to say, a pre-political—account of how these rights originate (Stilz 2019, 33). Stilz prefers an "occupancy" basis for territoriality, one rooted in the value of occupancy of a certain territory for individuals' lives (Stilz 2019, 35–36).

Yet, the reason that Stilz has some difficulty differentiating her own occupancy account of territorial rights from the proprietary one, going so far as to regard it as a "property-like" claim, is that it (as with the proprietary account) is fundamentally still an individualistic one. The reason for this is that Stilz dismisses "collectivist" accounts of territoriality such as those of David Miller and Margaret Moore. On the one hand, such accounts (for instance, Miller's) make unconvincing claims that national identities are a necessary basis for attachments to territory. On the other hand, they may (as with Moore) invoke "peoplehood" (i.e., political community) in a way that fails to explain the "value of collective self-determination" (Stilz 2019, 29).

I agree with Stilz about all this, and have made similar points earlier. But where does that leave her individualist account? Unfortunately, it is also unable to account for the *necessity* of particular "occupancy rights," since they are dependent on individuals *needing* to live in a certain way, in a certain place. But we know that *individuals* do not need to live in any particular place (territory) in order to survive, even to flourish. Human beings are uniquely malleable and adaptive. But human *communities* are not equally so. They *will* perish when separated from particular environments (or territories). This is a point indigenous peoples for instance frequently make (as emphasized in Chapter 2). Not only that: the *lands* will sometimes perish as well, when separated from peoples willing to sustain them—and above all, *protect* them—from other, more exploitative (i.e., extractive) practices in which different human populations may be interested.

All this is to say that in order to ground a robust and durable territorial right, a sovereignty claim—in fact, a claim of ecosovereignty—is needed in order to unequivocally incorporate the collectivist, and indeed political, dimension of territoriality. This brings us finally to the third justification for ecosovereignty—that it is the basis for the best kind of environmental governance. The last two chapters of this book will be devoted to explaining this in some detail, now that the argument for ecosovereignty as an interpretation of territorial rights and sovereignty claims is complete. In some ways, this argument for ecosovereignty as a type of environmental governance will recall some points made in the first two chapters, in which the claims to be made in this book were motivated as a conceptualization of ongoing political struggles against extractivism (and for environmental justice) in the present moment.

But a more complete argument must include two things: first, a statement of the goals of environmental governance, which will entail a certain understanding of what environmental justice involves (without necessarily giving a full theory of it), and second, a consideration of different kinds of environmental governance, and the reasons why ecosovereignty should be a basic principle underlying governance institutions that confront environmental problems. Since the point has been made repeatedly that territorial rights and sovereignty claims must be based on a certain view of the environmental crisis we now confront, it is important to give a more complete account of the values and principles necessary for facing this crisis.

This book is written from the perspective of a certain understanding of political ecology, a term more commonly used in geography and anthropology, than in philosophy. Nevertheless, the concept does usually involve an at least implicit view of environmental justice. This view may be briefly characterized here; it is the primary subject of the next chapter. Since political ecology as a form of inquiry constitutes an implicit challenge to political economy, it is generally taken to imply that any just distribution of social goods should include not just such things as property, capital, and income, but also natural goods such as soil, fresh water, biomass, and waste sinks. A complete conception of social justice should include an equitable distribution of these goods, as well as those produced by human labor.

Natural goods, since they are the material prerequisites of any "economic" goods with value-added, have historically been the object of considerable social conflict. So "ecological distribution conflicts" are those in which different human groups have engaged in order to procure natural goods (Martinez-Alier 2002). A just distribution of goods therefore entails an equitable resolution of such conflicts. Clearly, what is needed here is a more precise understanding of what "ecological equity" means. Another way to consider this is that, if the concept of "environmental justice" was originally developed to ensure an equitable distribution of environmental *harms*, its more complete elaboration must also include the equitable distribution of *benefits* from natural environments (Bullard 2018).

But what is the relation of ecosovereignty to such a view of environmental justice? If a conception of justice is formulated which is not matched by a form of governance that can realize that conception, then the conception itself must be questioned. This has been a problem, not only with certain views of economic justice but with those of environmental justice as well. For instance, various views of how to deal with anthropogenic climate change (i.e., "climate justice") have foundered on the manifest inability of existing political institutions to realize them (Jamieson 2014).

It has been a contention of this book that most political institutions—both "national" and "global"—do not have the legitimacy and/or efficacy to fully face the environmental crisis. Even worse, it is not clear that such institutions are really even motivated to do so, since they are themselves often complicit in the empowerment and enrichment of various political elites. There is accordingly a legitimation crisis that is affecting existing governance structures due to their evident failure to deal with social problems at a civilizational scale.

One thing that speaks in favor of a principle of ecosovereignty is its potential to empower peoples outside of existing institutions. Whether it is by facing problems of extractivism, ecocide, and resource capture through the politicization of struggles opposing such phenomena, or it is by legitimizing peoples identified by "spiritual" attachments to land and territory, ecosovereignty may enable resistance to environmental destruction in ways that escape the attention or capability of existing forms of governance. The full argument for why this is so will be given in Chapter 8. For now, it is important to begin consideration of why ecosovereignty is the best form of environmental governance by relating it to a particular idea of environmental justice; this is the subject of the next chapter.

References

Arendt, Hannah. 1948. *The Origins of Totalitarianism.* New York: Harcourt.

Arendt, Hannah. 1963. *On Revolution.* New York: Viking Press.

Arendt, Hannah. 1977. "What Is Freedom?" In *Between Past and Future: Eight Exercises in Political Thought.* New York: Penguin Books.

Banai, Ayelet, and Eszter Kollar. 2021. "Sovereignty and the Value of Self-Determination." In *Sovereignty as Value*, edited by Andre Santos Campos and Susana Cadilha. Lanham, MD: Rowman & Littlefield.

Buchanan, Allen. 1991. *Secession: The Morality of Political Divorce from Fort Sumter to Lithuania and Quebec.* Boulder, CO: Westview Press.

Buchanan, Allen. 1997. "Self-Determination, Secession, and the Rule of Law." In *The Morality of Nationalism*, edited by Robert McKim and Jeff McMahan. New York: Oxford University Press.

Bullard, Robert. 2018. *Dumping in Dixie: Race, Class, and Environmental Quality*. 3rd ed. New York: Routledge.

Cohen, Joshua. 2010. *Rousseau: A Free Community of Equals.* Oxford: Oxford University Press.

Dahbour, Omar. 2003. *Illusion of the Peoples: A Critique of National Self-Determination.* Lanham, MD: Lexington Books.

Dahbour, Omar. 2013. *Self-Determination Without Nationalism: A Theory of Postnational Sovereignty.* Philadelphia, PA: Temple University Press.

Dahbour, Omar. 2019. "On the Ecological Blindspot in the Territorial Rights Debate." *Territory, Politics, Governance* 7 (2): 217–32.

Davis, Mike. 2006. *Planet of Slums.* London: Verso.

Eckersley, Robyn. 2004. *The Green State: Rethinking Democracy and Sovereignty.* Cambridge, MA: Massachusetts Institute of Technology Press.

Elshtain, Jean Bethke. 2008. *Sovereignty: God, State, and Self.* New York: Basic Books.

Getachew, Adom. 2019. *World-Making After Empire: The Rise and Fall of Self-Determination.* Princeton, NJ: Princeton University Press.

Grimm, Dieter. 2015. *Sovereignty: The Origin and Future of a Political and Legal Concept*, translated by Belinda Cooper. New York: Columbia University Press.

Harris, Nigel. 1990. *National Liberation.* Reno: University of Nevada Press.

Hinsley, F. H. 1986. *Sovereignty.* 2nd ed. Cambridge: Cambridge University Press.

Jackson, Robert. 1990. *Quasi-States: Sovereignty, International Relations, and the Third World.* Cambridge: Cambridge University Press.

Jackson, Robert. 1999. "Sovereignty in World Politics: A Glance at the Conceptual and Historical Landscape." In *Sovereignty at the Millenium*, edited by Robert Jackson. Oxford: Blackwell.

Jamieson, Dale. 2014. *Reason in Dark Times: Why the Struggle Against Climate Change Failed—And What It Means for Our Future.* Oxford: Oxford University Press.

Koskenniemi, Martti. 2010. "Conclusion: Varieties of Sovereignty—Powers of a Paradox." In *Sovereignty in Fragments: The Past, Present and Future of a Contested Concept*, edited by Hent Kalmo and Quentin Skinner. Cambridge: Cambridge University Press.

Lipping, Jüri. 2010. "Sovereignty Beyond the State." In *Sovereignty in Fragments: The Past, Present and Future of a Contested Concept*, edited by Hent Kalmo and Quentin Skinner. Cambridge: Cambridge University Press.

Martinez-Alier, Joan. 2002. *The Environmentalism of the Poor: A Study in Ecological Conflicts and Valuation.* Cheltenham: Edwar Elgar.

Murphy, Alexander. 1996. "The Sovereign State System as Political-Territorial Ideal: Historical and Contemporary Considerations." In *State Sovereignty as Social Construct*, edited by Thomas J. Biersteker and Cynthia Weber. Cambridge: Cambridge University Press.

Pettit, Philip. 2023. *The State.* Princeton, NJ: Princeton University Press.

Philpott, Daniel. 2001. *Revolutions in Sovereignty: How Ideas Shaped Modern International Relations.* Princeton, NJ: Princeton University Press.

Rousseau, Jean-Jacques. 1978. *On the Social Contract*, translated by Judith R. Masters. New York: Saint Martin's Press.
Sachs, Wolfgang. 1999. *Planet Dialectics: Explorations in Environment and Development.* London: Zed Books.
Stilz, Anna. 2019. *Territorial Sovereignty: A Philosophical Exploration.* Oxford: Oxford University Press.
United Nations. 1960. "Declaration on the Granting of Independence to Colonial Countries and Peoples." UNGA 1514(XV). https://documents.un.org/doc/resolution/gen/nr0/152/88/pdf/nr015288.pdf
Vanderheiden, Steve. 2020. *Environmental Political Theory.* Cambridge: Polity Press.
Walzer, Michael. 1977. *Just and Unjust Wars.* New York: Basic Books.

PART IV

The Political Struggle

7

EXTRACTIVISM AND ENVIRONMENTAL JUSTICE

Extractivism, Sustainability, and Equity

Ecosovereignty is a political principle designed to counter the extractivist practices that are decimating natural environments, and peoples' livelihoods that directly depend on them. As such, it is a principle that legitimates political regimes in which peoples take control of ecosystems within which they live, in order to preserve their natural goods and services, both for themselves and for their nonhuman inhabitants. As we will see in the next chapter, it can legitimate certain political laws and institutions that make it more possible for peoples to defend—and in some cases, take or recover—ecosystems endangered by extractivism.

Nevertheless, the ecosovereignty principle is not explicitly formulated to criticize the justifications of extractivism, nor to propose a comprehensive alternative to it. Such concepts as "progress," "development," and "growth" have been used to justify the ongoing extraction of "natural capital," and have been subjected to intensive examination and critique for this reason. At the same time, different attempts to formulate an alternative—usually under the rubric of "environmental justice"—have been made, with varying degrees of coherence. While these are not the main subject of this book, it is important to see how ecosovereignty expresses a certain view of environmental justice (EJ, hereafter) that may not be a familiar one, even for those who have some acquaintance with theories of justice as they have been applied to environmental problems.

The reason for this lack of familiarity is that the critique of extractivism is not often a central or explicit part of most conceptions of EJ. This is the case for various reasons, as we will see shortly. To develop such a critique that does

DOI: 10.4324/9781003568155-12

eventuate in a conception of justice requires, first, a social theory that rejects the dominant notions of social progress and development in industrial capitalism. It also requires, second, the formulation of an alternative to the ideology of progress that has legitimated extractivism as necessary for the continued functioning of industrial societies. This chapter will give brief consideration to a group of concepts that culminate in a critique of extractivism as a conception of EJ.

A social theory that rejects progressivism and developmentalism begins with a certain understanding of the natural economy, and the *natural capital* that is the basis of real wealth. It connects this understanding with a conception of *ecological sustainability* that rejects the necessity of capital accumulation and the extraction of natural goods that it requires. The result of this understanding is a notion of *ecological distributive justice* (EDJ) that is distinct both from standard versions of (economic) distributive justice and from non-distributive conceptions of EJ. The core concept of EDJ is that of *ecological equity*, which names the goal that claims of ecosovereignty seek to realize.

But the initial step in outlining these ideas that underlie the ecosovereignty principle is to consider the puzzling absence of extractivism from much of the discussion of EJ, at least until quite recently. This will motivate the attempt to formulate a conception of EJ that centers on the extractivist processes that are a fundamental cause of the environmental crisis in this century.

The Concept of Environmental Justice

The term, "environmental justice," as is now well-known, originated in the USA in the 1980s to describe the equitable distribution of waste disposal and other environmental hazards. From this beginning, the term has subsequently been applied in an increasing number of ways to various environmental problems and their equitable solutions. Nevertheless, it is rare for the concept, even now, to be applied to the problem of resource extraction, except under the rubric of "climate justice," where it designates equitable distributions of the costs of climate mitigation and adaptation. Realizing that, as Anna Willow puts it, "extractivism transforms 'nature' into an instrument of injustice," will, in turn, require transforming our conception of environmental justice in important ways (Willow 2019, 17). The fact that this realization has been largely absent from various conceptions of EJ until quite recently may seem inexplicable. But it is not, if the origins, and subsequent development, of the concept are kept in mind.

In a brief history of the idea, Steve Vanderheiden identifies three generations of thought about EJ. The first was concerned with "injunctive remedies" to end inequitable practices, especially those of waste disposal among poor and minority populations in the USA (Vanderheiden 2016, 324; see especially Bullard 2018). The second generation continued the focus on the disposal of effluents, toxics, and similar harms, but introduced a concept of "restorative" justice, principally by way of compensation for past damages. The third generation focused on more global, than local, environmental harms, and sought international cooperation of various kinds to remedy them. Once this shift occurred, the concept was increasingly used to designate equitable solutions to anthropogenic climate change—as a planetary-wide problem of waste management. With this emphasis, the term, "climate justice" came into general usage.

These three stages in the development of the concept of EJ exemplify changes in thinking about remedies to environmental damages that have already occurred. In different ways, they involve halting, repairing, or compensating for environmental harms of different kinds. The remedies taken tend to involve either legal decisions, financial awards, or both, as injunctive, reparative, or compensatory actions for damages or harms. But the most significant fact about the versions of EJ is that they concern primarily, if not exclusively, harms done to environments, not goods (or services) derived from environments. In that sense, justice involves not alternative uses, or perhaps conservation, of various goods (land, water, and so on), but corrections (or compensations) for previous destructive activities.

Nevertheless, the concept of justice (in its various guises) developed to deal with damage to environments is, from an analytical rather than historical perspective, fundamentally a *distributive* concept. This is because the compensations given, or the reparative actions taken, are usually thought to be monetary, if only because this is the easiest way to deal with the costs and damages resulting from environmental damage. As we will see, this is but one type of justice, viewed from a theoretical perspective. Yet, what is remarkable about this application of distributive justice is that the just distribution itself is not one of environmental goods and services, or even of (intact) environments, but of *economic* goods that are compensatory for the harms done to ecosystems. That the goods and services themselves are not what is (re)distributed is perhaps due to the fact that they are regarded as having already been degraded or destroyed. This is just as much the case for "climate justice" as it is for cases of "discriminatory" pollution or other environmental degradation. In the former case, debates about just distribution usually concern the allocation of costs for mitigating global warming (e.g., reducing fossil fuel usage) or for adapting to changing environmental conditions as a result of such warming (Vanderheiden 2008).

Before distinguishing distributive from other forms of justice as applied to environments, it is important to be clear about how the distributive concept works as it has been applied here, from its origins to recent applications to climate issues. First, the concept is largely "economic" inasmuch as reparations or compensations are usually monetary. As we will see, this is not the only good that can be distributed, though it is the usual one in most theories of (distributive) justice. Second, the concept involves a focus on harms, since the goal is not the (equitable) distribution of goods (e.g., land), but the more equitable distribution of harms or damages (e.g., pollution), and/or financial allocations that compensate for their effects.

In this respect, "environmental" justice as it is usually understood is really more a matter of reparations or compensations for losses, than it is a matter of distributing equitably the benefits (goods and services) derived from natural environments. Here we have the first indication of why extractivism is not necessarily a (central) issue for theories of EJ. An assumption may be operating that regards environmental harms (e.g., pollution) as inevitable, at least to some degree. What is at issue in considerations of justice is therefore the (equitable) distribution of the negative effects of the (industrial) processes that generate these harms. Applied to extraction of natural goods, there may be a similar assumption: such goods—turned into inputs ("resources") for industrial processes—will be extracted (e.g., mined, fished, farmed, and harvested) from *somewhere*. Injustice occurs when they are only extracted from vulnerable communities or environments, while leaving other similarly resource-rich places untouched or preserved.

One way in which this assumption may be operating is in the conflation of "extraction" with "extractivism." For instance, David Schlosberg, writing about what distributive equity means in EJ, states that "indigenous communities face *disproportionate impacts* due to the environmental degradation of development and resource extraction" (Schlosberg 2007, 84). What is interesting about this statement is that the injustice lies not in the extraction itself, but in its "disproportionate impacts" on, for example, indigenous peoples. What is the difference between extraction of resources and extractivism (as an industrial-capitalist practice)? Here is what Anna Willow says: "Unlike extraction, extractivism is relentless in its quest to obtain more than required for simple sustenance . . . . Extractivism values natural resources . . . for the profits they yield" (Willow 2019, 17).

It is unclear whether extractivist practices are even a concern for most theorizations of EJ. For instance, a comparison could be made between the practices subject to standard understandings of EJ and those characteristic of extractivism. Gordon Walker, in his 2012 survey of concepts of EJ, lists the following problems:

waste disposal siting
air quality control
flood risk vulnerability
urban greenspace availability
anthropogenic climate change mitigation and adaptation

(Walker 2012, vi–vii)

By comparison, in her 2019 study of extractivism, Willow lists the following practices as extractivist:

forest logging
river damming (for both water and power)
mining for industrial uses (including energy production)
coal mining (as a special case)
oil and gas extraction

(Willow 2019, vii–ix)

These lists need not be taken as exhaustive (there are a number of other activities that could qualify for one or the other). But they do provide an interesting contrast. An obvious difference between these two lists is that the second concerns the causes of environmental destruction, while the first concerns the effects of such practices. Most EJ theorists view these effects as matters of distribution, which can be done justly or unjustly, equitably or inequitably. But what about the causes of such effects, rooted in the extraction for profit of natural goods? Certainly, some would view such practices as more or less justified, though in some cases subject to limitations in order to preserve some natural environments. But in what way might these extractivist practices themselves be regarded as unjust? Only if they are viewed as *particular* (and problematic) uses of environmental goods and services could they be contrasted with different possible uses and therefore be regarded as potentially unjust. In other words, the practices might be regarded as unjust *if* they deprive some peoples of environmental goods and services necessary for maintaining their livelihoods, while they serve others' consumer preferences.

A distributive concept of EJ, focused on the equitable distribution of the effects (i.e., harms) of (natural) resource use, is therefore not a good basis for opposing extractivism. The reasons are that, first, such a concept focuses on effects rather than causes of environmental destruction, and second, that it regards just distributions largely as the result of economic (i.e., monetary) compensation given to rectify these harmful effects. By the time the environmental effects of resource extraction have occurred, the damage has already been done—and, in many cases, may be nonrectifiable, at least for the peoples originally affected in the extraction zones. Monetary compensations obviously do not prevent environmental harms from occurring in the first place; nor can they necessarily provide the means for restoration of impacted ecosystems, or of the livelihoods that have been lost.

It should, however, be understood that distributive EJ is only one approach to EJ, and that others should be considered possible bases for criticizing extractivism. Two in particular that could be considered are participatory and recognitive conceptions of EJ. The participatory concept has been given extended treatment by Kristin Shrader-Frechette, who defines it as the attempt to "remove the unjust constraints that some people have over other people's lives and actions." This attempt seeks to "help ensure that there are institutional and procedural norms that guarantee all people equal opportunity for consideration in decision-making" (Shrader-Frechette 2002, 28). Shrader-Frechette applies this to environmental injustices by advocating a "Principle of Prima Facie Political Equality," (PPFPE) an idea that of course can also be applied in other contexts (Shrader-Frechette 2002, 24–27). She argues, "One way to achieve participative justice in environment-related decision-making is to follow the PPFPE, according to which stakeholder and expert deliberation are given equal weight" (Shrader-Frechette 2002, 28).

Would such an approach be effective as a counterweight to extractivism? This is unlikely, whatever virtues the concept has when used to deal with harms resulting from environmental damage. In cases involving resource extraction, the chief conflict is not, as with cases of pollution and similar harms, likely to be between "stakeholders" (e.g., victims) and experts. It is more likely to be between *different* (and different kinds of) stakeholders. One of the major problems with stakeholder theory—and the PPFPE seems to be a version of this—is assessing the relative weight of different stakeholder claims. Giving them equal weight is not a way to resolve such conflicts justly. The different kinds of stakeholders—and the different uses to which they might put resources (i.e., environmental goods)—cannot be assumed, absent specific argumentation—to be equal. For instance, local users, distant consumers, industrial corporations, and government agencies may all be stakeholders. What is the reason why their claims, based on variable needs, interests, and desires, should be given equal weight? The PPFPE "presumes that equality is defensible and that only different or unequal treatment requires justification" (Shrader-Frechette 2002, 27). But this presumption only seems warranted when the different purposes of interested parties are roughly comparable (as in the paradigmatic case of citizens' equal rights to political participation). What could warrant the presumption in cases of extractivism, when the purposes of resource use are so disparate?

Another approach to defining the nature of EJ without reference to distribution is that of recognitive justice. The idea, which originated in the work of such philosophers as Charles Taylor, Iris Young, and Axel Honneth, has been applied to EJ more recently by David Schlosberg, Kyle Whyte, and others. Part of the connection resulted from the fact that, as Whyte observes, there was "case after case in which citizens of the US and other countries were more likely to face environmental risks owing to their race, ethnicity, gender and class" (Whyte 2017, 116).

The general idea of recognitive justice is that distributive injustices often have their origin and maintain their effects as a result of the lack of respect accorded to members of distinct social groups. This means that justice must contain, as an essential aspect, respect for and recognition of social and cultural differences, rather than simple affirmations of "strict equality" (Whyte 2017, 113). In the realm of EJ, exclusive focus on distributive equity results in failing to acknowledge that, as Schlosberg puts it, "there is a crucial link between a lack of recognition and the inequitable distribution of environmental bads; it is a general lack of value of the poor and people of color that leads to this distributional inequity" (Schlosberg 2007, 60).

The crucial question about this recognitive approach, however, is not whether it can account for the origins of certain patterns of environmental harms, but whether it can constitute an effective critique of extractivist practices that lead to these harms. The *siting* of extractivism in, for example, Native American lands can certainly be traced to settler colonial ideologies that denigrate native peoples. But does the lack of recognition of indigenous cultures provide a basis for the critique of extractivism itself? If, for instance, such extractivist enterprises were located in lands settled by Europeans (as in some oil and gas "fracking" in North America), would this be a just distribution of the practice? Or, if the damming of rivers, or logging of forests, is conducted by native-owned companies, rather than foreign ones (as in various regions of South Asia), would this too constitute a just distribution of the harms?

These quandaries derive from a crucial ambiguity in the meaning of "difference." As Whyte states, "Differences can be cultural, including language, or can involve land- or water-based practices associated with people's subsistence and flourishing" (Whyte 2017, 119). In either case, "The central concern of many environmental justice groups is community and cultural *survival* in a system where recognition is denied and communities and cultures are thoroughly devalued" (Schlosberg 2007, 62; italics added). But survival in these two cases may mean very different things. In the first case, where recognition involves respect for diverse cultures, languages, and customs, it can take various forms—including economic support for schools, cultural practices, and *relocation* to more secure places in which such practices can continue, away from the resource sacrifice zones where extraction is occurring. In the second case, it involves according ownership, territorial, and sovereignty rights to peoples so that they can *remain in place*, surviving on the lands and waters in which their cultures (in the second sense) have meaning.

The first case of recognitive justice allows the extinction of relations with lands and waters, which, as was discussed in Chapter 2, are a fundamental basis of indigeneity; the second case preserves this basis and opposes industrial practices that seek to turn such environments into "resources." But this just is to say that focusing on the recognition of difference is to miss the crucial feature of extractivism—not its obliteration of cultures as artifacts, but its destruction of biophysical environments within which cultures retain their living significance.

To prevent the latter from occurring, it is not sufficient to "recognize" cultural differences. Peoples must be *empowered* to hold the line against activities that seek to turn nature into a source of economic value, above all else. But to do this, nature must be (re)distributed to those capable of, and committed to, such a fight. What this means for a revised concept of EJ is the matter to which we now turn.

Natural Capital and Ecological Distribution Conflicts

None of this is to say that EJ is concerned solely with distribution. Other aspects of justice are certainly relevant here. As Joan Martinez-Alier, who has discussed EJ extensively from the standpoint of political ecology, writes: "social mobilizations over resource extraction, environmental degradation, or waste disposal are not only about the distribution of environmental benefits and costs . . . *they are also about participation in decision-making and recognition of group identities*" (Martinez-Alier 2016, 556). Many of the considerations leading to the principle of ecosovereignty discussed earlier—the sovereignty concept itself, the ideas of indigenous rights, livelihood, and territoriality—suggest additional dimensions to the struggle against extractivism. Perhaps "justice" is not even the best rubric under which to understand this struggle: (collective) rights or (political) membership might serve instead. But, for the moment, EJ continues to be the discourse within which critiques of extractivism can most easily be understood. Philosophy can aim to reform language, but it must also accommodate its concepts to some degree to the historically derived meanings that inform current understandings. For now, justice remains an irreducible context for thinking about the environmental crisis.

To the extent that this is the case, we must attempt to use such terms as justice in a context-specific way to refer to what is actually at issue. It is in this sense that the concept of distribution remains relevant, though not in the way that it is often understood. As previously mentioned, when distribution is thought to be at the heart of EJ, it often means some form of economic or monetary compensation for environmental harms. As Richard Schroeder writes, "Typically, distributive mechanisms involve royalty payments or taxes of some sort, which are either funneled into community development programs or into state coffers, from which they are theoretically disbursed in efforts to offset conservation costs" (Schroeder 2000, 54).

When EJ is thought about in this way, it minimizes other aspects that are sometimes captured by the concepts of participatory or recognitive justice, such as claims based on livelihoods or indigeneity. The distribution of economic compensation for resource extractions, for instance, can sometimes actually exacerbate the inequities resulting from the capture of natural wealth. In return for certain economic benefits being conferred, peoples may suffer "displacement, dispossession, insult, injury, harassment, and outright terror" (Schroeder 2000,

63). When EJ is restricted to distributive redress for harms, it can ignore a "wide range of fundamental claims to resources, place-based identities, and livelihoods" (Schroeder 2000, 64).

But is distributive justice necessarily identified only with this restrictive conception? This is a moment in which philosophy can benefit from taking a wider view of certain concepts, such as justice. From the perspective of those interested in political ecology—geographers, anthropologists, and others in the social sciences—environmental justice tends to include some of the aspects of what philosophers regard as non-distributive concerns. For instance, the concept of "ecological distribution conflicts" (EDCs), as developed by the ecological economist, Joan Martinez-Alier, refers to many different types of disputes and conflicts over environmental goods and services (Martinez-Alier 2002). He, and others who have used this concept, often regard the resolution of such conflicts as equivalent to the achievement of EJ (Martinez-Alier 2023). Such resolutions do not necessarily take a monetary form, and often include changes in land tenure, recognition of rights to use, conservation of ecosystem goods, and restrictions on resource extractions. Understanding this very different conception of distribution should lead us to formulate an alternative to economic distribution, that is to say, EDJ. This idea requires some explication, both in terms of the nature of the goods to be distributed and of the ways in which such distributions are just or unjust.

All societies—or so political ecologists maintain—are founded on a relation to nature, and depend for their reproduction on the availability of certain natural goods. But some societies "plunder" or consume such goods without much thought for the long-term consequences or the necessity of restraints on use. Such societies dominate, rather than co-exist with, nature (Horkheimer and Adorno 2002). In particular, industrial capitalist societies—through a combination of regimes of capital accumulation and the use of industrial methods of production—have such a relation to nature. Given the need to maintain the socoecological foundations of life, this relation of domination would seem to be antithetical to the maintenance of socioecological systems *in general* (Leff 2021). Despite the attempt to link industrial capitalist societies with a development paradigm that can avoid conflicts over increasingly scarce resources, such conflicts continue to grow around the world, and are increasingly eventuating in poorer, dysfunctional versions of such societies (Crary 2022, 20–43).

Political ecology, originating in geographical research on agrarian peoples and environments, has expanded to become an interdisciplinary study of conflicts over environmental goods and services in countries and regions around the world (Robbins 2012). Any study of conflict has an evaluative component, especially if the conflict is over matters relating to the very survival of the peoples involved. The fact that peoples' cultures and survival is often at stake has led to increasing interest in a critique of development theory; and this critique is now an important part of political ecology itself. The idea of (economic) development is often countered by espousal of values of sustainability and sufficiency, rather than those of growth and efficiency. Similarly, environmental justice has itself taken on a particular meaning distinct from the origins of the concept in differential assessments of environmental harms. This meaning, sometimes designated as that of ecological equity, concerns the distribution of environmental benefits, as much as that of harms.

Environmental justice, from the perspective of political ecology, therefore concerns primarily the equitable resolution of EDCs. These conflicts occur when different societies, states, or organizations seek to capture as much of a (natural) resource as possible, as a basis for current use, and possible future growth. The fact that these conflicts are increasingly occurring in the twenty-first century is a reminder that the "political economy" continues, as throughout human history, to depend on a "natural economy." This natural economy consists, not of the products of human labor, but of the material goods, or "throughput," necessary for economic production to occur in the first place. These material goods—"natural capital"—are taken "for free" from natural environments. Without them, a political economy that produces "value-added" (i.e., by human labor) would not be possible.

In the current period, increasing scarcities of natural capital are resulting from the depletion of the natural economy that supplies these material throughputs (Daly 1996, 78–79). These are not absolute scarcities of "natural resources" in most cases. But they are significant enough to affect the availability and pricing of material assets. What makes this situation so dire—and can lead to EDCs resulting in unjust distributions of natural, and therefore productive, wealth—is that scarcities of natural capital are not susceptible to redress through more egalitarian redistributions of wealth. This is because natural scarcities are the result of limits on the *total* resources available for human flourishing and even survival.

It is important to note that EDCs are often the result of *different* ways of utilizing natural capital. Two major forms of processing material throughputs are, on the one hand, efforts at efficient production in industrial capitalist economies, and on the other hand, attempts to use *and conserve* natural capital through "cyclical" production processes. The focus on efficient production of goods in industrial economies can be regarded as a way of ignoring the limits to material throughputs (Daly 1996, 33–34). The result is, as Wolfgang Sachs writes, a "throughput economy where enormous quantities of natural inputs are finally transformed into enormous quantities of . . . waste" (W. Sachs 1999, 176).

One response to this problem of depletion and waste is a focus on "dematerializing" the industrial economy (Sachs, Loske, and Linz 1998, 114–15). On some accounts, industrial economies are sustainable *if* they can succeed in replicating natural systems' conservation of resources. Ways to do this include more efficiently using existing natural materials, thereby minimizing the loss of such materials, especially nonrenewable ones, and substituting industrially produced materials for ones found in natural environments. Certainly, there have been some successes, both with more efficient use of natural goods, and with "substitutability." But can industrial economies *as a whole* de-materialize in this way? Problems abound; for instance, the increasing pressure to produce more that arises both from continual population growth and from ever rising consumer demands may overwhelm efficiency gains. At the same time, as methods of, for example, energy production change from fossil-fuel-based to renewable, the type of material inputs changes, but not necessarily the quantity. Solar and wind energy are already absorbing increasing amounts of land, building materials, and minerals. Overall, dematerialization cannot be counted upon to maintain industrial economies, once the natural goods upon which they depend become scarce (Martinez-Alier 2002, 252–53).

But an alternative form of resource use that seeks to exist within the limits of the natural economy may be able to utilize natural capital in ways that do not deplete the resource. The point is to recognize, as J. B. Callicott writes, that the "human economy is a subset of the economy of nature" (Callicott 2016, 6). This is particularly essential for many of "the rural poor in the South [for whom] biological resources . . . are often the only or the most important means of survival" (Sachs, Loske, and Linz 1998, 78).

This approach to both utilizing and conserving natural capital involves a cyclical conception of the production process that uses natural materials, and a solar (rather than carbon)-based system of energy production. The idea of a cyclical production process suggests that natural materials can be re-cycled in such a way as to replenish the biomass that is the source of these materials. An example of how to do this is the maintenance of agricultural polycultures, rather than the monocultures (usually dedicated to raising crops for profit, rather than subsistence) that lead to soil depletion (Altieri 1987). Another example is the "industrial ecology" approach of re-cycling used industrial products, rather than regarding them as waste matter (Callicott 2016, 6–7). In addition, the natural economy-oriented approach cultivates still available biomass in those parts of the world that possess "megadiversity"—the tropics and especially fertile parts of the temperate zones (Martinez-Alier 2002, 2).

Natural capital is, in short, the object of EDCs, whether or not it is utilized by industrial or "natural" forms of production. Environmental justice, from the political ecology perspective that views natural capital as the primary environmental good, means the just distribution of such material goods. One environmental philosopher who has seen this connection is Andrew Dobson, through his concept of "critical natural capital." In his view, natural capital is nature, regarded as "'an economic asset of some description' [Brian Barry]" (Dobson 1998, 40). *Critical* natural capital is, then, natural capital that is "critical to the production and reproduction of human life" (Dobson 1998, 43).

Two questions arise about this idea of just distribution: [1] to whom is natural capital to be distributed, and [2] by what principle(s) is its distribution to be justified? Considering the first question, Dobson puts considerable weight on intergenerational justice, a concern that does not *directly* connect to our present interest in a critique of extractivism. It does, however, connect to the extent that, as Dobson argues, (ecological) sustainability is a primary consideration in determining just distributions. If "ecosystem health" is, as he argues, a fundamental feature of sustaining natural capital, then only distributions that maintain the integrity of "ecological processes" can be considered just (Dobson 1998, 44).

Yet, Dobson does not seem to consider sufficiently the ways in which sustainability may require redistributions of natural capital to certain peoples in the present (as opposed to distributions that benefit future generations). His major comment about this is indecisive: "We might try . . . arguing that doing justice by the present poor will coincide with doing justice to future generations but, once again, it is by no means obvious that the two *do* coincide" (Dobson 1998, 256). This may be true if "the present poor" are simply people with low incomes. But if they are considered those who, as political ecologists have pointed out, are pursuing an "environmentalism of the poor," that is another story (Guha and Martinez-Alier 1997).

Often, as pointed out earlier, it is those in the (Global) South (and some in the North) that are dependent on natural capital for their survival who have the greatest stake in maintaining the ecosystems upon which they depend. In this sense, it is those who were referred to in prior chapters as "ecosystem peoples," as well as those, often from urban areas, seeking land or other natural goods and services in order to achieve more food and other security, who may be most deserving of ecological redistributions. To show the relation of such distributive claims to extractivism, however, we need to consider at somewhat greater length the different environmental policies that could be brought to bear to provide natural capital to peoples engaged in EDCs.

Justice and Sustainability

Understanding the relation of extractivism to various concepts of environmental justice involves comparing two divergent notions of how to solve problems of economic and environmental maldistribution. On the one hand, advocates of "sustainable development" argue for a set of policies that use intensive means of exploiting natural resources in order to achieve widespread (economic) development of poor and disadvantaged peoples and groups. On the other hand, advocates of "ecological sustainability" maintain that only the preservation of ecosystems and natural goods and services, both for the currently poor, and for future generations, can address such problems. While the latter approach opposes extractivism as a general matter, the former seeks to comprehensively utilize all natural goods in the service of the world's population as a whole. Whether this involves resource extractions in so-called sacrifice zones is a matter of debate. The implications of these two approaches for justly resolving EDCs will be discussed in the following section.

Sustainability, then, has importantly different meanings. It has sometimes been thought to be about maintaining a full range of options available within a given ecosystem (Norton 2005). "Transcending" the "limits" of an ecosystem means *reducing* the options available to those living "in place." This is because *not* living in place requires consuming goods from multiple environments; and, at scale (i.e., for a large population), this form of industrial production and mass consumption will create pressure to transform many ecosystems into agricultural monocultures—that is, plantation economies. For those remaining within the natural contexts of discrete places, such places will offer only an impoverished and unsustainable set of "natural" goods that can survive *after* a transition to industrial agriculture.

Industrial capitalism—despite the ideology of material progress that accompanies it—actually results in a narrower and more impoverished way of life in many respects. This is because, first, industrial consumers are cut off from direct exposure to the pleasures and wonders of the natural world which persons in agrarian societies often take for granted (MacKinnon 2013). Second, the need to provide basic needs satisfactions for many hundreds of millions of additional workers and consumers (as population increases with the industrial production of necessities) results in an inevitable standardization of needs (Soper 2020). By sustaining a living ecosystem, as Bryan Norton emphasizes, the multiple options offered by such an environment will remain available to future generations—and this is indeed a large part of what sustainability is usually taken to mean: "[It] is *a relationship between generations such that the earlier generations fulfill their individual wants and needs so as not to destroy, or close off, important and valued options for future generations*" (Norton 2005, 363, italics in original).

While this is certainly an essential aspect of the meaning of sustainability, it is also important to note that such a loss of options can occur *within a generation*, as well as across them (Martinez-Alier 2016, 556). The more general definition that Norton gives of sustainability is therefore worth emphasizing: "Living sustainably is *maintaining* the mix of important options [for living]; living unsustainably is losing them, narrowing the range of options" (Norton 2005, 364). Sustainability, then, whether understood in relation to present or future generations, means the maintenance of natural capital, rather than its depletion for short-term consumption or profit.

Achieving sustainability requires a change in the nature of the motivations embedded within forms of social reproduction (such as family life, schooling, and cultural institutions). Rather than viewing efficiency—and its frequent accompaniment, (economic) growth—as the parameter for success, some conceptions of sustainability focus on the value of "sufficiency" instead (Princen 2005). From this standpoint, it is acceptable to view ecosystems as possessing resources, materials, and services for human use *if* this use is regarded as limited to the sufficient and equitable distribution of such ecological goods. Sufficiency can be defined as the state that is attained when a society's "outputs" (products and services) are measured by, as Sachs puts it, "the quality of the civilization it creates *out of limited means*" (W. Sachs 1999, 182; italics added). That "quality" is the equitable distribution of essential benefits and services, maintained over medium to long time scales (i.e., > two generations), and capable of resilience in the face of major environmental shocks and disturbances.

Even an ecologically much more *efficient* society may outrun its means if it does not strive for sufficiency in this sense. Without a self-limitation of this sort—a rejection of goals and purposes that require for their realization more than the existing resources of the ecosystem within which the society exists—nothing will be gained by increasing the ecological efficiency of the economy. The human economy's "expansive dynamics will undermine all successes achieved from boosting resource efficiency" (W. Sachs 1999, 185).

The problem with aiming at efficiency, rather than sufficiency, even when ecologically sound methods are used, is the lack of resilience should there be either an "overshoot" of resources, or unanticipated challenges. Even while aiming at greater economic efficiency, communities cannot reasonably assume that they have prepared themselves for adversities arising from resource depletion or environmental change. Resilience will not result if resource use does not accept limits to use *well below* what ecosystems can supply.

The reason for this lies in the insufficiency of human knowledge, at least when it comes to planning for environmental contingencies arising from unanticipated overuse or depletion of resources. There is no well-founded scientific method for studying the limits of ecosystem services and products, when unspecified human uses are taken into account (Holling, Berkes, and Folke 1998, 351). Resilience—understood as the capacity for communities to deal with unanticipated and rapid disturbances of ecological equilibrium—cannot be guaranteed, if all resources are in use or already allocated. This is because there are no adequate "feedback mechanisms" that can generate data useful for responding to such disturbances (Holling, Berkes, and Folke 1998, 355).

As long as societies attempt to live on the "margins" of ecological viability, the danger that such margins will be exceeded—either by constantly expanding human desires, or by sudden environmental disruptions—will be everpresent (Holling, Berkes, and Folke 1998, 351). A standard example of this has to do with fisheries. The "ecoefficient" approach—that of determining the "maximum sustainable yield" that a given fishing ground can bear—has been shown repeatedly to result in overfishing and catastrophic depletion of fish stocks. The traditional approach, geared to harvesting *sufficient* fish stocks to adequately feed a limited, local population, did not have this result. By limiting *how* fishing could be done—restricting or prohibiting certain types of boats, nets, and so on—fish stocks were historically maintained for long periods without risking depletion (Holling, Berkes, and Folke 1998, 357).

In the present global environment, such a way of regarding social reproduction is especially important, and still very present, in those communities and among those peoples who are clearly dependent on natural systems for their survival. This is the case, above all, for hundreds of millions of people in Southern countries who live and work in rural environments. For them, the conservation of nature, and the diverse lifeforms that it contains, "constitutes an indispensable aspect of social and economic development" (Sachs, Loske, and Linz 1998, 78).

From a different, "developmental," perspective, the fact that so many people continue to live in rural settings, dependent on their immediate environs for the satisfaction of basic needs is itself the problem. When "ecological efficiency" becomes a goal, due to concern for "global" poverty and "underdevelopment," a comprehensive modernization of the economy is regarded as the right way to deal with the depletion of resources. This efficiency-oriented approach has been variously described as one of "sustainable development" or "ecological modernization" (J. Sachs 2015; ecomodernism.org 2015). The focus lies on achieving much greater efficiency in the use of natural resources, since such resources are in increasingly scarce supply. From this perspective, natural goods are important only to the extent that substitutes cannot be found for the "environmental services" that are necessary for the maintenance of human life. But to the extent that substitutions can be found for natural materials, or greater efficiencies achieved in their use, maintaining a connection to natural systems has no other value. Furthermore, from the ecomodernist perspective, it may be *disadvantageous*, because it implies the immutability of peoples' relations to places, in a world in which greater human populations may require *more* mobility due to labor market fluctuations and increased residential concentrations in cities.

One crucial difference between these approaches to sustainability is geographical or spatial. Ecological sustainability, which emphasizes the local scale, tends to be more sustainable and resilient than development practices where the scale of resource use is more global (i.e., used by consumers who do not live in the ecosystem where the natural goods are found). The reason for this difference is simple: locally based socioecological systems contain "feedback loops" that allow for checks on resource use that are unsustainable: "local social systems of rights and responsibilities develop for any resource deemed important for a community . . . rules arise and evolve according to local needs" (Berkes and Folke 1998, 7).

Conversely, when communities do *not* manage local natural goods by regulating human use, the state of such goods suffers accordingly: "[If] communities are not involved in the active management of their natural resources, they will use resources destructively" (Agrawal 2000, 39). There are a number of ways in which this can happen. When states, for instance, rather than local communities, take control of natural goods, they may do several things: use them for their own purposes, accord usage rights to neighboring communities previously uninvolved in resource management, or sell resource rights to business interests located in cities or even overseas.

It follows that intensification of development regimes (e.g., through more efficient use of environmental goods and services) can encourage further overuse of resources. Yet, that intensification is what the idea of development entails: the replacement of local socioecological systems with an industrial capitalist system on a potentially global scale. The idea of sustainable development therefore implies an increasing simplification of means for the exploitation of resources, and these means suggest the possibility of the continuing expansion of resource exploitation (i.e., through extractivism). As Wolfgang Sachs writes,

> The development discourse is deeply imbued with Western certainties like progress, growth, market integration, consumption and universal needs, all notions that are part of the problem, not the solution . . . they preclude the search for societies that live graciously within their means.
>
> *(W. Sachs 1999, 28)*

The ecological sustainability approach need not preclude all concepts of development or modernization. For instance, Herman Daly distinguishes between development and growth, such that development constitutes *qualitative* improvements in social reproduction, while growth is equivalent to *quantitative* increases in economic production. Daly's view is that development can occur without growth, and this is a legitimate aim for sustainable socioecological systems (Daly 1996, 69). Similarly, Joan Martinez-Alier advocates an "alternative modernization" that utilizes locally sustainable systems but attempts to scale them up to deal with the problems of resource scarcity more globally (Martinez-Alier 2002, 207–08).

It is important to note that the developmental conception of sustainability is also aspirational, in that it relies on the *hope* for technological breakthroughs that can make continued resource efficiencies feasible. As Jeffrey Sachs puts it, "*if* we are clever and apply ourselves to the study and design of new sustainable business practices and technologies, sustainable development is both feasible and affordable" (J. Sachs 2015, 43; italics added). This view makes extractivism part of a larger strategy for meeting the increasing demands of an evergrowing global population. Another part of the strategy—and one that makes extractivism seem an acceptable option—is the idea of "infinite substitutability between human-made and natural capital" (Dobson 1998, 78). As Dobson comments, it may be doubted whether, once this part of the approach is considered, the idea of sustainable development is really about the conservation of environmental goods and services at all.

Ecological Equity

It is now time to bring the different strands of thinking about environmental justice together, and see how they can help to justify an alternative to extractivism. The concept of *ecological equity* will be used here to do this (Gadgil 2000). Ecological equity (EE) is equivalent to the just resolution of EDCs, such that natural goods are distributed to those who can maintain a regime of ecological sustainability in the present and across several generations. EE is a principle of EDJ, as opposed to the concept of *economic* distributive justice that has been the brunt of various criticisms, such as those discussed earlier.

What is distributed is access to, ownership of, rights to, or sovereignty over (as the case may be) natural goods and services found in ecosystems. (Which of these manifestations of EE is relevant depends on the contrast class of claimants—direct or indirect users (e.g., extractivist agencies or consumers), property owners, territorial states, or sovereign entities.) It is not sufficient for economic or monetary equivalents to these goods to be distributed (e.g., other goods with equivalent values). The goods themselves must be equitably distributed. To *whom* they are distributed are peoples attached to, directly using, and capable of maintaining long-term, ecosystems—that is, ecosystem peoples. The *limits* on such distributions are set by the functional requirements of sustainability.

Some further comments on this conception may be useful. EE means, more concretely, access of subsistence producers to the natural goods and services necessary for their continued survival *in their place of residence*. It also means the adoption of forms of human-natural interaction that preserve the wealth of the ecosystems on which life *in these places* depends. It is crucial to note that EE involves a "right to place" for peoples embedded within ecosystems (and perhaps even for peoples who *seek* a self-sufficient life within them) (Anguelovski and Martinez-Alier 2014). Why do such peoples have a right to place, rather than a right to the goods that come from these places? One reason is that EE concerns the distribution of natural goods themselves, not the proceeds, benefits, or wealth that come from such goods. Were the principle to be satisfied by *equivalents* of natural goods, then there would be nothing inherently ecological about the concept of equity. It would just denote a quantity of benefits or goods that is fungible in relation to other such quantities. This would be to turn a *natural* good into a *market* good (Polanyi 1944). What this does is to make such goods inessential—exchangeable for equivalents, substitutes, or comparative benefits. But nature is not inessential; it is the basis for all goods, and indeed for all life. This realization begins (and also ends) in the recognition of the importance of place(s) for human and nonhuman life. A right to place is just as essential for protecting ecosystems as a claim to the environmental goods and services found in these places.

Clearly, EE requires an approach to ecosystems that emphasizes ecological sustainability over sustainable development. Since the rationale for this conception of equity is the direct access to natural goods, developing these goods through resource extraction and efficient use risks their diminution or disappearance. Furthermore, the limits on their use are generally set by the needs of the *direct* users present in the ecosystem, along with provisions made for future generations of such users. Other uses (such as trade for nonlocal goods) are permissible as long as they do not degrade or diminish the sustainability of the socioecological system as a whole. Should users of a resource seek to turn it into a vehicle for accruing profit and economic capital, they would violate the EE principle and become complicit in unjust distributions of the resource.

This conception of EE contributes to the critique of extractivism by justifying distributions of natural capital to those who directly use them, rather than to corporations, states, or other agencies seeking to profit from, monetize, or market natural goods and services. Resource extractions used for any purpose other than peoples' livelihoods is prima facie unjust, on this account. This means, among other things, that when there is an EDC involving differing uses for a resource, the claimants seeking to use it for purposes of self-sufficiency would have priority over other claims. In this regard, EE has similarities to the principle of just distribution understood as an "entitlement." Dobson, following David Miller, distinguishes three bases for principles of (distributive) justice—"need, desert, and entitlement" (Dobson 1998, 76).

The connection of natural goods with peoples having a prima facie claim to them is based on these peoples' integration into existing ecosystems. So entitlement to claim priority in the distribution of these goods follows from this connection. It is different from the proprietary connection that is often assumed to be essential for an entitlement claim, as in Robert Nozick's conception of justice (Nozick 1974; Dobson 1998, 77–78). Here, it is the *dependence* on natural goods that creates an entitlement, not a structure of private ownership. Yet, this is neither a relation of desert, with peoples deserving of a certain distribution due to their efforts, nor a relation of need, even though needs may enter the relation at some point. Needs is a notoriously malleable concept, and can be asserted by many parties, and for different sorts of uses. For instance, distant consumers may assert a need for goods that require resource extractions of various kinds. While some of these needs may be genuine, basic, and/or survival needs, it still is inequitable to assert their primacy over the livelihood requirements of local users in particular ecosystems, who may be limited to accessing the goods and services of that ecosystem. Needs of distant consumers, served by long-distance supply chains, can be served by various sources, some of them local to the consumers themselves.

Extractivism is therefore best challenged as unjust by a conception of environmental justice that first, focuses on the distribution of natural (not economic) capital, and second, accords priority of access to those who are committed to the sustainability of these natural goods. This is because other distributions (of resources as economically valuable), and other priorities (of global consumers and their purported needs) will not provide strong reasons not to engage in resource extractions. For instance, one such reason, compelling on various grounds, is the amelioration of the poverty of distant consumers. Achieving (ecological) equity and justice involves acknowledging the entitlement to natural goods that peoples closely connected to ecosystems have the right to claim as an entitlement, despite different claims that might seem justifiable for other reasons.

The brief consideration of an example of extractivism may serve to show how equity considerations would work, but also to exhibit some of the complexities in such cases. One complexity is that the case is historical; but I will bracket the matter of "historical justice" for now, because it introduces elements that are not present in most contemporary cases of extractivism. Considering an historical case, however, does give us the ability to study the long-term impacts of extractivism. The case is that of the damming of the Colorado River in the Desert Southwest region of North America (already mentioned in Chapter 4). Creating what was, at the time of its opening in 1935, the world's largest dam (the Hoover Dam) enabled agricultural and energy companies to access the water and hydroelectric power of a river in order to build an "oasis civilization" in a region (southern California) previously unconnected to the watershed of the Colorado River (Fradkin 1981, xviii). At the same time, it meant the death of one of the largest river deltas in North America, where the Colorado had emptied into the Gulf of California (between the Baja California peninsula and mainland Mexico).

Today, the river no longer reaches the Gulf of California, having destroyed the habitat and livelihoods of most of the indigenous Cocopa people who had lived in the delta region (Willow 2019, 70–72). By contrast, the cities of Los Angeles, Phoenix, San Diego, Las Vegas, and Tucson would either not exist at all, or as much smaller towns without the water and power extracted from the Colorado. Of course, these cities were a fraction of their size today prior to 1935, when large-scale extractions of river water and power began. Their development meant substantial financial and political rewards for those who had initiated and managed the extractions.

Putting aside the possibility of reparations for historical injustices arising from this case, in what respect is this to be regarded as an "ecological distributive *in*justice," violating the principle of equity in the distribution of environmental goods associated with the Colorado River? Somewhat surprisingly, there is a forum for consideration of this question, since the U.S. states involved in managing the river hold talks regularly about the proper distribution of the river's resources. Such talks include states with no necessary connection to the river (California), states that govern small parts of the river's watershed (Nevada), and states that are more centrally connected to the river itself (Arizona, Colorado, and Utah). In addition, the damming of the river has not precluded the periodic release of waters to the original delta, releases that have dramatic short-term positive impacts on its still-extant riverine ecology (Willow 2019, 77).

The question to ask about such a case is whether drastically revising the agreements regarding the use of the river's resources is necessary to achieve ecological equity. There are further complications. Most of the original inhabitants of the river delta have left. In addition, recent climatic changes resulting in reduced snowmelt in the mountain ranges that feed the river have themselves drastically curtailed the water flow and therefore already induced political maneuvering over the distribution of remaining resources. The main determinant of equity in this case, however, is not water quantity or political power; it is the ecological character of the river and its watershed. As Willow writes, "Ecologically speaking, water is transported from where it is supposed to be to where it does not belong, forced to follow a sociopolitical logic rather than [the] fundamental laws of ecology" (Willow 2019, 73).

As a first step toward equity, then, a redistribution of existing benefits away from regions outside the river's watershed to those inside would seem to be mandated. A second step would be to reduce draws on the water sufficiently to enable the river to reach the delta (and the gulf) again. This does not amount to "pulling the plug" on the forty million or so inhabitants of southwestern cities. But it does involve dramatically raising the *price* of water and power in that region, sufficiently to induce lower levels of extraction and use. As with previous episodes of water use in the desert region, this will undoubtedly have effects on habitation and agricultural usage in the area. Managing those effects will be politically challenging, since the extractions have gone on for almost a hundred years, enabling the hydraulic culture of the region. But the reality is that, due to the effects of "global heating" on the area, change is coming anyway. Will it be equitable; or will it further unjust distributions of what remains of the natural goods available?

In some ways, the case just discussed is not a typical one, if there are any that are typical. Its historically important role makes it to some extent sui generis. Furthermore, the fact that the primary beneficiaries are not far "out of area" (the Colorado River basin), but adjacent to it makes it seem exceptional. In fact, there is a history of river damming in the southwestern region, though never before over such great distances, or at such large scales. It is often the case with water extractions that the conflicts occur between adjacent users. For instance, the Narmada River case in India concerned the attempt of downriver farmers to dam the river, depriving upriver farmers of their lands (as was previously discussed in Chapter 3). Other forms of extractivism can involve long supply-chains, where resources are transported across the globe to (very) distant industries and consumers.

In cases where extractions involve mining or removal of biomass (e.g., lumber) from the extraction zones, there is less often a close connection, as in the aforementioned examples involving water rights, between the sacrifice zones (where the extractions take place), and particular areas of usage. Often it is the case that the natural goods that are plundered from these zones are processed industrially for consumption in many different regions. In these cases, perhaps the more common ones, halting extractions would have immediate impacts on certain industries, but not necessarily on particular groups of consumers, as in the riverine cases. In that regard, the previous example of extractivism is a "hard case," compared to those where the issue is more immediately one of corporate profits versus local peoples' livelihoods. In these instances, halting extractions would have immediate benefits for the ecosystem peoples, and only delayed effects on distant consumers. Nevertheless, there is no doubt that reducing extractions of coal, oil, natural gas, lumber, and various minerals would greatly impact industrial societies in the long-term.

The Colorado River case, however, partly due to its historically unprecedented (and well-publicized) character, clearly shows how the basis of any extractivist projects involve: [1] distant users, [2] destructive changes to local ecosystems (and the peoples that inhabit them), and [3] the application of political power and economic wealth to favor the former and ignore the latter. Equity in these cases must involve redressing the imbalance between [1] and [2]; and to do this requires the use of political power to counterpose to that employed to facilitate extractivism in the first place.

Of course, beyond this case, and indeed, aside from extractivism as such, environmental justice is a large concept, and involves other considerations as well—not least, the claims of future generations. But the degradation and decimation of ecosystems that is proceeding through intensive resource extractions must first be halted before the natural wealth that is left can be distributed equitably. This is in fact the purpose, not only of the concept of EDJ but also of the suite of concepts presented in this book related to the ecosovereignty principle—food and livelihood sovereignty, indigenous and place-based rights, and ecosystem-based territorial rights and self-determination claims.

Ecological Distributive Justice, Ecosovereignty, and Governance

This is to say that, while EDJ is the normative heart of a solution to the problem of extractivism, understood as a matter of injustice, it is essential to employ other concepts that can mobilize political power toward this end. This is why environmental justice, more generally conceived, not only is a distributive concept, but also involves other matters, such as the recognition of (ecosystem and indigenous) peoples, and claims to political self-determination by such peoples.

The relation between environmental justice and the ecosovereignty principle also needs brief consideration. As a general matter, it is an instance of the relation between the goals of political action—in this instance, the achievement of ecological equity—and the means and justification of political action. Ecosovereignty, in other words, is a concept that pertains not to justice per se, but to governance and authority—the political means for achieving communal goals. Yet, in another sense, ecosovereignty is a feature of a "political" principle of justice; it is necessary to provide the ideas and institutions necessary for the achievement of EE (Rawls 2001, 181–82). Peoples who have, as a matter of justice, priority in the distribution of environmental goods must also have the means to claim, and thereafter to protect, those goods. This is what ecosovereignty is designed to justify.

Furthermore, inasmuch as extractivism is a process involving states, corporations, and in some instances, global agencies as well, it can only be successfully opposed by peoples who have justifiably claimed their rights to the goods and services, lands and waters, territories, and environments that they require in order to live a sustainable existence. Ecosovereignty is a principle designed to give such peoples a sense of the justice and legitimacy of these claims. Certainly, extractivism will not be ended by the very actors that have perpetrated it in the first place.

As long as (environmental) justice is regarded as a matter of the proper distribution of fungible goods and wealth, the environments upon which so many peoples depend will continue to be regarded as so much material for exploitation and extraction. Justice must be rethought as the equitable distribution of the natural goods themselves, not the resources or wealth to be gained from their despoliation. This is probably the only way that the scales (of justice, of power) will be righted to the advantage of the most disadvantaged peoples in the world today—as well as of the wounded living natures that are also struggling to survive.

What remains is to consider the forms of environmental governance that would be best suited to embodying the ecosovereignty principle, and the idea of justice which the principle seeks to establish and protect. This is the subject of Chapter 8.

References

Agrawal, Arun. 2000. "'Community' and Natural Resource Conservation." In *Nature, Production, Power: Towards an Ecological Political Economy*, edited by Fred P. Gale and R. Michael O'Gonigle. Cheltenham: Edward Elgar.

Altieri, Miguel. 1987. *Agroecology: The Scientific Basis of Alternative Agriculture.* Boulder, CO: Westview Press.

Anguelovski, I., and J. Martinez-Alier. 2014. "The 'Environmentalism of the Poor' Revisited: Territory and Place in Disconnected Global Struggles." *Ecological Economics* 102: 167–76.
Berkes, Fikret, and Carl Folke. 1998. "Linking Social and Ecological Systems for Resilience and Sustainability." In *Linking Social and Ecological Systems: Management Practices and Social Mechanisms for Building Resilience*, edited by Fikret Berkes and Carl Folke. Cambridge: Cambridge University Press.
Bullard, Robert. 2018. *Dumping in Dixie: Race, Class, and Environmental Quality*. 3rd ed. New York: Routledge.
Callicott, J. B. 2016. "Ecological Sustainability." In *Routledge Handbook of the Philosophy of Biodiversity*, edited by Justin Garson, Anya Plutynski, and Sahotra Sarkar. London: Routledge.
Crary, Jonathan. 2022. *Scorched Earth: Beyond the Digital Age to a Post-Capitalist World*. London: Verso.
Daly, Herman. 1996. *Beyond Growth: The Economics of Sustainable Development*. Boston, MA: Beacon Press.
Dobson, Andrew. 1998. *Justice and the Environment: Conceptions of Environmental Sustainability and Dimensions of Social Justice*. Oxford: Oxford University Press.
Ecomodernism.org. 2015. "An Ecomodernist Manifesto" (online).
Fradkin, Philip. 1981. *A River No More: The Colorado River and the West*. New York: Knopf.
Gadgil, Madhav. 2000. "Ecology and Equity." In *The Use and Abuse of Nature*, edited by Madhav Gadgil and Ramachandra Guha. Delhi: Oxford University Press.
Guha, Ramachandra, and J. Martinez-Alier. 1997. "The Environmentalism of the Poor." In *Varieties of Environmentalism: Essays North and South*. London: Earthscan.
Holling, C. S., Fikret Berkes, and Carl Folke. 1998. "Science, Sustainability, and Resource Management." In *Linking Social and Ecological Systems: Management Practices and Social Mechanisms for Building Resilience*, edited by Fikret Berkes and Carl Folke. Cambridge: Cambridge University Press.
Horkheimer, Max, and Theodor Adorno. 2002. *Dialectic of Enlightenment: Philosophical Fragments*, translated by Edmund Jephcott. Stanford, CA: Stanford University Press.
Leff, Enrique. 2021. *Political Ecology: Deconstructing Capital and Territorializing Life*. New York: Palgrave Macmillan.
MacKinnon, C. B. 2013. *The Once and Future World: Nature as It Was, as It Is, as It Could Be*. Boston, MA: Houghton Mifflin Harcourt.
Martinez-Alier, Joan. 2002. *The Environmentalism of the Poor: A Study in Ecological Conflicts and Valuation*. Cheltenham: Edward Elgar.
Martinez-Alier, Joan. 2016. "Global Environmental Justice and the Environmentalism of the Poor." In *The Oxford Handbook of Environmental Political Theory*, edited by Teena Gabrielson, Cheryl Hall, John M. Meyer, and David Schlosberg. Oxford: Oxford University Press.
Martinez-Alier, Joan. 2023. *Land, Water, Air, and Freedom: The Making of World Movements for Environmental Justice*. Cheltenham: Edward Elgar.
Norton, Bryan. 2005. *Sustainability: A Philosophy of Adaptive Ecosystem Management*. Chicago, IL: University of Chicago Press.
Nozick, Robert. 1974. *Anarchy, State, and Utopia*. New York: Basic Books.
Polanyi, Karl. 1944. *The Great Transformation*. Boston, MA: Beacon Press.
Princen, Thomas. 2005. *The Logic of Sufficiency*. Cambridge, MA: Massachusetts Institute of Technology Press.
Rawls, John. 2001. *Justice as Fairness: A Restatement*. Cambridge, MA: Belknap Press.
Robbins, Paul. 2012. *Political Ecology*. 2nd ed. Chichester: Wiley-Blackwell.

Sachs, Jeffrey. 2015. *The Age of Sustainable Development.* New York: Columbia University Press.
Sachs, Wolfgang. 1999. *Planet Dialectics: Explorations in Environment and Development.* London: Zed Books.
Sachs, Wolfgang, Reinhard Loske, and Manfred Linz. 1998. *Greening the North: A Post-Industrial Blueprint for Ecology and Equity*, translated by Timothy Nevill. London: Zed Books.
Schlosberg, David. 2007. *Defining Environmental Justice: Theories, Movements, and Nature.* Oxford: Oxford University Press.
Schroeder, Richard. 2000. "Beyond Distributive Justice: Resource Extraction and Environmental Justice in the Tropics." In *People, Plants, and Justice: The Politics of Nature Conservation*, edited by Charles Zerner. New York: Columbia University Press.
Shrader-Frechette, Kristin. 2002. *Environmental Justice: Creating Equality, Reclaiming Democracy.* Oxford: Oxford University Press.
Soper, Kate. 2020. *Post-Growth Living: For an Alternative Hedonism.* London: Verso.
Vanderheiden, Steve. 2008. *Atmospheric Justice: A Political Theory of Climate Change.* Oxford: Oxford University Press.
Vanderheiden, Steve. 2016. "Environmental and Climate Justice." In *The Oxford Handbook of Environmental Political Theory*, edited by Teena Gabrielson, Cheryl Hall, John M. Meyer, and David Schlosberg. Oxford: Oxford University Press.
Walker, Gordon. 2012. *Environmental Justice: Concepts, Evidence and Politics.* London: Routledge.
Whyte, Kyle. 2017. "The Recognition Paradigm of Environmental Injustice." In *Routledge Handbook of Environmental Justice*, edited by Ryan Holifield, Jayajit Chakraborty, and Gordon Walker. London: Routledge.
Willow, Anna. 2019. *Understanding Extractivism: Culture and Power in Natural Resource Disputes.* London: Routledge.

8

ENVIRONMENTAL GOVERNANCE AND ECOSOVEREIGNTY

Extractivism and Environmental Governance

Achieving environmental justice, especially by opposing extractivism, also requires a system of environmental governance. This system should include both normative principles justifying political rights and responsibilities, and institutions that are committed to realizing these principles. The concepts of environmental justice and governance could therefore work in tandem: justice requiring governance for its realization; governance requiring justice in order to legitimate its political authority. The principle of ecosovereignty justified in this book serves the purpose of providing a fundamental justification of (ultimate) authority that can legitimate environmental governance institutions. In this chapter, we will consider several possible forms that such institutions could take.

Standard conceptions of governance involve certain ethical principles (usually, of justice or rights) combined with institutions that can realize these principles by pragmatic accommodation to existing conditions and institutions (Swart and Perry 2007). But there is another way to justify forms of governance. From a more realist point of view (as expressed, for instance, by philosophers such as Hannah Arendt), political institutions can embody specifically political values, such as those of sovereignty or self-determination. Even John Rawls, usually considered a moralist, claimed in his later work, that his theory of justice was "not applied moral philosophy," but pertained to the "domain of the political," and constituted "political values: they arise in virtue of certain special features of the political relationship" (Rawls 2001, 181–82).

In terms of environmental governance, attention needs to be paid to the norms implicit in different scales and systems of governance. The question should be

DOI: 10.4324/9781003568155-13

asked: Given the norms implicit in these forms of governance, which is best suited to the achievement of environmental justice, and the ending of extractivist practices? Enunciating a principle of ecosovereignty as a definition of the ultimate authority obtaining in environmental governance institutions is an important means of achieving environmental justice.

The focus of the discussion here will be a comparison between different justifications and instantiations of environmentally responsible states and of common property institutions. Most discussions of environmental governance, however, make a different comparison. While ignoring the commons as a possible governance institution, they focus on how to understand states as capable of effective environmental governance. But, in addition, states are often compared with (possible) global governance institutions as a means of dealing with environmental problems. So this section will begin with a few preliminary comments about the possibility of global environmental governance. In addition, it is important to remember that much of the discussion about environmental governance generally does *not* concern, or even mention, extractivism as a core part of the problem of environmental destruction. Instead, most considerations of governance tend to concern how to deal with anthropogenic climate change, or with biodiversity loss. Regarding environmental governance from the perspective of ending extractivism will therefore provide a unique angle on the legitimacy and viability of various proposed governance institutions.

At several points in this book, particularly in Chapters 3 and 5, the cosmopolitan or global approach to environmental crisis has been considered—and rejected. This is consonant with a recent re-emphasis on *local*, and often state-based, solutions to environmental destruction. Nevertheless, it is important to recall why global solutions to problems of extractivism, climate change, and biodiversity collapse are not likely or, in some cases, even desirable. Attempts to deal with extractivism and other environmental problems have often been driven by two very different intuitions about the scale of environmental problems and their solutions. On the one hand, environmental *problems* have often been seen as affecting large, even planetary, scales. Certainly, many environmentalists would argue that such problems as global warming operate beyond the scale of the locality, country, or even continent. They have a global reach. On the other hand, environmentalist *solutions* often have a local, or at least ecosystem-specific, scale. As Robert Goodin has put it, "If there is anything truly distinctive about green politics . . . it must surely be its emphasis on decentralization" (Goodin 1992, 147).

In terms of the matter of governance, global solutions, as was pointed out in Chapter 5, are difficult to operationalize, for various reasons. Probably the most important one is that, in the absence, of truly hegemonic global institutions, global governance will continue to rely on the power, and be subject to the imperatives, of powerful states. These states will often have little reason to transform or limit their own extractive enterprises, or the growth-oriented consumption patterns that drive them.

To the extent that global governance does become a real possibility, it will almost certainly—as has been the case so far—be based on institutions with doubtful, if any, political legitimacy. This is because such governing bodies will at best be established indirectly by states acting in the name of, but not directly responsible to, their own citizenrics. This combination of the improbable effectiveness of global institutions, along with their questionable legitimacy, suggests that solutions will lie on a more local scale.

There is an additional "motivational" problem with thinking of governance on a global scale. Can global governance institutions rely on a sufficiently robust form of political identity and loyalty to support attempts to limit or end extractivism? This form of political identity is generally thought of as cosmopolitan. Cosmopolitanism has been defined as the view that persons are, *ethically,* members of a human—that is to say, planetary—community, rather than of particular communities, such as countries, regions, or ecosystems (Caney 2016, 239). An entailment of this view is that the moral rights and duties that persons possess by virtue of this common humanity should obtain (equally) to all persons (globally).

Yet, as discussed in Chapter 3, such a philosophy can only weakly validate ideas of territorial rights, or other locally specific laws or policies. This is because such initiatives tend to favor the interests and needs of local peoples, not all persons equally. States generally recognize that this is the basis of *their* legitimacy; but global institutions do not have this option. They can appeal only to generalizable interests in, e.g., human rights or global justice.

In terms of human rights, there have been attempts to employ the concept to specify and condemn environmental harms of various kinds (Woods 2016). What such attempts usually focus on are certain material preconditions of (human) well-being, such as clean water (or air), basic health, and the bases for meeting subsistence needs (e.g., food, housing). In all such cases, however, the purported *human* right to such things is one that all persons presumably possess. Nothing in this view justifies a right to protect a specific ecosystem from resource extractions. It is here that the problem of "excessive generality" in global human rights doctrines becomes manifest.

The problem may be understood in the following way. Were there a right of all persons to one or another precondition of well-being, such a right would have to be realized in such a way that all persons could obtain that particular good. Does this leave room for some to have more of a good, or perhaps a specific instantiation of a good, as a result of living in a particular place (and in a particular way)? The right—to the extent that it is one that all humans may claim—cannot be to a specific instantiation of a good, only to *an instance* of the good. Such goods may be accessed in one place; but that place (and that person) have no priority over the right to such a good by all, *somewhere.* This problem of generality means that human rights *may* serve as the basis for a critique of extractivism only inasmuch as it prevents its victims from realizing their rights (e.g., to subsistence, to air and water) *at all*. If they can be realized in another way, in another place, then there would be no violation of human rights as such.

Global governance has also been linked with a specific cosmopolitan idea of global justice. The idea of global governance is justified by a conception of global justice rooted in the idea of a (global) social contract. There are some problematic assumptions built into such an idea—for instance, the concept of a global society, and what that could mean (Dahbour 2019, 33–37). Nevertheless, assuming for the moment that such an idea can be suitably justified, there is still a problem in applying this notion of global justice to extractivism, ecocide, and other environmental harms at the local scale. Essential to the notion of a global social contract is the idea that the relation between the location of persons and the location of natural resources on the planet is "morally arbitrary" (Beitz 1999). Whether or not persons are located in proximity to the resources necessary to satisfy their needs is of no ethical significance, on this view. It is simply a geographical fact that certain persons are located nearer to natural resources than others, and from the point of view of contractarian social theory, it is regarded as arbitrary.

But from a different perspective, this supposed arbitrariness is a blindspot in the contractarian theory of global justice. Connections between peoples and ecosystems (lands, waters) are only arbitrary from a standpoint in which individual persons are regarded abstractly as rational, moral agents. From a realist perspective, the geographical and historical origins of peoples' environments and livelihoods are at least partially determinant of the normative content of those lives and worlds. It is in part the material origins and character of peoples' environments that *gives value* to those things, for those people.

Another way of putting this is that the very idea of "resources" is value-laden. A natural environment such as an ecosystem can be given a description based on its geographical, ecological, and other characteristics. But this does not mean that it inherently contains materials that are of use to humans in certain ways. Only the valuation of these materials, based on human purposes and goals, transforms natural features into natural resources. In other words, natural features are "natural kinds," resources are "intentional kinds," so designated by humans on the basis of particular intentions (for their use) (Kolers 2012; Crist 2019, 246). But who is doing the designating? If distant others designate certain natural features as resources for (their) use, does this have equivalent—or even greater—normative status than the designation by local peoples of these features as being either resources for their (perhaps different) uses, or non-resources with other (e.g., symbolic) significance? An argument needs to be made here for why this would be the case; claiming that local peoples' attributions of significance are "arbitrary" betrays a bias against non-instrumental valuation, where nature is not viewed necessarily as constituting resources for human purposes. If the case is one of local versus non-local usage, then the bias is a "globalist" one in the sense that those who have the material wealth and technological ability to pursue their interests on a planetary scale are presumed to take priority over those whose interests are rooted in localities.

The reduction of these human-natural relations to the status of arbitrary facts makes it hard to be critical of the expropriation by others of these resources, as long as they are used to satisfy some persons' needs (or at least, *perceived* needs) somewhere else. The result of such expropriations may be hardship and loss for peoples actually living in diverse and plentiful ecosystems. But, from a globalist point of view, compensations, resettlements, and other forms of redress are sufficient responses to the problem. In fact, the replacement of such ecosystems with, for example, more efficient agricultural systems, may be justified if they result in the provision of *more* goods and services (i.e., for global consumers). On this account, extractivism would not be unjust, provided that compensation for loss occurs and resources are used more efficiently to satisfy the needs of others.

This view has been thoroughly criticized, both earlier in this book and elsewhere. These criticisms eventuate in a view that, though a global perspective on the environmental crisis is essential for highlighting certain aspects of the problems, solutions—especially to extractivism—will have to be tailored to local needs and circumstances. There are, however, ways in which this may be less true about certain environmental problems other than extractivism—anthropogenic climate change, or species extinctions in certain world regions.

For instance, the mitigation of global warming will require efforts in a great number of different countries and regions, in order to reduce the use of fossil fuels (Jamieson 2014). But adaptation is different: distinct manifestations of climate change will almost certainly require differential adaptations to changing climatic conditions. Nevertheless, it is fair to think that the climate problem needs address, in some manner, beyond the local scale (Gardner 2013). Similarly, some aspects of the problem of species extinctions and biodiversity loss escapes purely local solutions. Areas such as polar or deep ocean environments, which lack significant human populations, are affected by long-distance human interventions to exploit natural goods. Accordingly, measures by many states and communities, in some coordinated fashion, may need to be taken in order to protect such environments from human exploitation, as has happened with some success in Antarctica already. For most natural habitats, however, the real issue is one of how specific forms of human-natural interactions do, or do not, negatively impact such habitats. For this, global governance seems less well-suited to dealing with the problem.

To achieve environmental justice, especially as characterized in Chapter 7 as ecological equity in the distribution of natural goods and services, global institutions are unsuitable. Because conditions under which extractivism is operative vary widely, the solutions will have to be tailored to, and achieved by, the (ecosystem) peoples most affected. To do this, only governance institutions—states or commons—working at the local scale will be effective.

Despite the problems with global governance, this does not mean that global institutions—in particular, legal institutions such as the International Criminal Court—cannot be of some help in combatting extractivism. While the practice itself has yet to take on legal ramifications at the global level, there has been a concerted effort to criminalize ecocidal actions by businesses and governments (ecocidelaw.com). To the extent that resource extractions contribute to ecocide—and it is surely one of the main causes—actions can be pursued in interstate arenas to limit its scope, and punish its perpetrators. Certainly, more traditional diplomatic agreements could also help to limit the sway of extractivists, of course provided that states were willing to cooperate in this manner.

In the end, global governance institutions will not be able to materially prevent or discourage extractivism, since their efficacy is clearly dependent on state and local actions, both to motivate the institutions in the first place, and to enact the needed changes on the local scale. The upshot is that we must look elsewhere—either to states or to commons—for institutions that can limit extractivism and achieve a modicum of environmental justice for peoples dependent on ecosystems for their livelihood needs.

"Green" Democracy and/or Liberalism

Not surprisingly, the state has been the primary locus of discussions about environmental governance. States of various kinds have had to deal with the management of environmental problems for some time now—the problems themselves have become inescapable. Furthermore, though there are instances of so-called market solutions to such problems, more often markets have been the *cause*, or at least a contributing factor, to such problems. The impetus of markets to capitalize goods of all sorts, including environmental goods—turning nature into "natural resources"—has meant that limitations on this process must come from elsewhere. States have sometimes exercised restraint on market-driven imperatives to capitalize and commodify everything. Furthermore, the absence of effective global governance institutions, as well as the atrophy or destruction of commons institutions in the past two hundred years, has meant that states often constitute the chief counterweight in many countries and regions to environmental despoliation for economic and market-oriented uses.

Yet, from another perspective, states seem remarkably ill-suited to the role of exercising and enforcing restraints on environmental destruction. States are often committed to goods inimical to environmental rationality. Two, in particular, come to mind: political aggrandizement and economic growth. While states are not *only* in the business of territorial expansion and the war-making that this often requires, this is certainly one enduring function of state institutions generally. Growth, or "development," in the current jargon, is also not always the business of states. But it frequently is, for the reason that the benefits—in terms of the expansion of state revenues—are so often manifest to rulers and regimes.

Nevertheless, despite these obvious problems, it is also unclear to many what means for achieving environmental restraint and equilibrium are available if states are *not* employed for this purpose. As flawed as they may be, states still seem an obvious mechanism for achieving environmental goals. For this reason, there has arisen a substantial literature in environmental philosophy concerning how states can be employed in this way. Much of this literature concerns how to modify or otherwise use central features of state-oriented ideologies and institutions to make them into "ecostates" (as they are sometimes called). Three aspects of this discussion will be examined here. First, the ideological bases for modern states in democratic and/or liberal ideas will be examined for the possibility that these theories of state legitimacy can become "green." That is to say the possibility that democratic and/or liberal institutions can become oriented to environmental, and possibly anti-extractivist ideas, in addition to their commitments to more core ideals, will be considered. Second, some proposals for changing the functional roles of states as welfare-enhancing or rights-enforcing institutions so that they can include environmentalist goals will be examined. In particular, it will be considered whether extractivism can be excluded from its historical role in funding and enabling state power. Third, the idea that the environmental crisis has reached a stage in which "emergency" powers need to be employed by states in order to deal with the effects of the crisis will be entertained. This is an idea expressed by the earliest advocates of ecologically responsible states in the 1970s, as well as by some recent responses to the burgeoning environmental crisis at the present moment.

In turning to the idea of a "green" democracy, it is important to understand how democracy, as a justification of state power, is to be understood. For some theorists interested in modifying democratic theory for environmentalist purposes, democracy is best understood in terms of rights. The idea seems to be that claims to environmental sustainability could be instantiated as the "democratic right" of citizens to survival in a healthy environment. In this way, environmental goods could be "built in" to democratic governance. As Robyn Eckersley notes, "pressing environmental claims as rights is intended to make such claims non-negotiable—or at least less negotiable than they currently are" (Eckersley 1996, 216). In other words, a "democratic right" is intended to take the object of such rights out of the hands of unruly democratic majorities who may, or may not, otherwise endorse them. So the notion of democratic rights rejects or seeks to limit a democratic majoritarianism.

The deeper problem here is this: democracy is a procedural principle of political legitimacy: decisions are legitimate if they were validated by a majority of the citizens of a state. Democracy *implies* majoritarianism, as Rousseau was the first to recognize: what the majority thinks is right *is* right. But environmental sustainability is a substantive good, not a principle for determining valid political authority. Sustainability, in other words, is good (if it is) whether or not democratic majorities decide that it is. There is very little democratic about *that*. While democratic states *may* institute policies aimed at protecting environmental goods, there is nothing about democracy that suggests that they necessarily *will* do so.

Furthermore, there is a well-known problem of democracy that is specifically applicable to the critique of extractivism. This is the problem of majority rule over minorities. Many cases of extractivism affect primarily minority populations of states—often, though by no means only, indigenous populations. In these cases, it may well be a fact that majorities—including democratic majorities—of the citizenry are quite happy to exploit the resources of certain ecosystems, as long as they affect only a minority of the population. In such events, a strictly democratic system might even hasten the demise of fragile ecosystems, if democratic majorities seeking cheap food, fuel, and other goods have no particular stake in maintaining those environments.

This is a critique of "green" democracy that William Ophuls, in his pathbreaking book from the 1970s, *Ecology and the Politics of Scarcity*, first pointed out (as we will discuss more next). Do democratic theorists have an answer? Robyn Eckersley, in her 2004 book, *The Green State*, attempts one: using the example of the "precautionary principle" that prioritizes consideration of environmental damages even when scientific evidence is not conclusive, she writes that, in a democratic system, it should not be possible to ignore such a principle. The way to do this is to "constitutionally entrench the precautionary principle in the same way that basic democratic rights are constitutionally entrenched" (Eckersley 2004, 135). Despite her subsequent claim that this would not place the principle "beyond the reach of democratic debate," it seems that this is precisely the purpose of constitutionalizing it in the first place. While a critique of extractivism founded on principles of the protection of ecosystems and ecosystem peoples dependent upon them is not *inconsistent* with democratic principles, it remains unclear how it could be guaranteed, or even facilitated, by them, in societies where majorities are wedded to industrialized life and consumption patterns.

Problems with the liberal theory of the state are comparable to, but somewhat different from, those with democratic majoritarianism. Liberalism justifies state power on the basis of individual rights, rather than on particular decision procedures, as in democratic theory. Traditional liberal theory, as formulated by John Stuart Mill and some other nineteenth-century philosophers such as T. H. Green, asserted that the highest value of the free individual would best be achieved through the establishment of certain institutions, such as a legal system that ensured basic liberties, and a political system that encouraged individual citizens to cultivate the virtues of free thinking and self-development. Liberalism, in this traditional form, therefore constitutes a specific political culture. This culture, as is evident in Mill's work, *could* be consistent with certain ideas that are central to modern environmentalism, such as the notion of a steady-state (i.e., non-growth-oriented) economy (Mill 2004). Furthermore, traditional liberalism's emphasis on securing the conditions for free self-development *could* involve, in Piers Stephen's words, "countenancing increased state regulation, economic planning, and redistribution" (Stephens 2016, 58). A conception of environmental justice might be entailed by this view.

An implication of this type of liberal theory is that the conditions of free self-development would prohibit destruction of ecologically intact societies or social groups that would allow such self-development. But there is a problem with the view. What if the culture of peoples resident in particular ecosystems does *not* recognize the value of some liberal principles—for instance, freedom of thought or belief, or other individual liberties? Presumably such societies would forfeit the right to claim priority or protection over the socioecological conditions of their lives. In other words, if liberalism is a *specific* set of cultural practices and social institutions, then it would not justify the protection of peoples living in nonliberal ways. Certain indigenous peoples, for instance, might not recognize individual rights to self-development that are aspects of liberal morality. Would they accordingly "forfeit" their *collective* rights to living in place on a particular territory? While there might be different answers to this question, it seems clear that the "liberal" answer would be "yes." It may be that other entitlements *should* be given up so that persons within such societies have the freedom of exit from a nonliberal culture.

Problems of this sort led key liberals in the twentieth century to formulate a different variety of liberalism, one that was more centered on the toleration of distinct ways of life. A key figure in this version of liberalism was John Rawls, who, in his book, *Political Liberalism*, formulated a theory of the state dedicated to the toleration of different conceptions of the good life that individuals might choose. This was to be made possible by establishing a framework of liberal institutions based on Rawls' well-known principles of justice (Rawls 1993). The standard way in which this form of liberal theory is characterized is that it is "neutral" between different conceptions of the good (life). This neutrality idea, and the version of liberalism based upon it, has become hegemonic in twenty-first-century liberalism, even though it raises difficulties for the application of that theory to environmental problems (Stephens 2016, 59).

A few liberal theorists, such as Marcel Wissenburg, have tried to incorporate some environmental commitments into their versions of liberalism. Yet, they come up against the neutrality problem. This problem may be most succinctly defined as the inability of liberal theory to either mandate environmental preservationist policies or to prohibit ecosystem destruction *if* this runs up against the individual preferences of citizens of liberal societies. Only if these citizens themselves clearly *prefer* such policies can the liberal state enact them. So Wissenburg argues that "individuals [should] get their fairest possible chance to accomplish their plans of life, whatever their plans may be" (Wissenburg 1998, 168). This *may* involve preservation of ecosystems, if this is a basis for giving individuals this chance. Wissenburg formulates a second-order principle (after more basic liberal principles) designed to do this. This "restraint principle" states, "No goods shall be destroyed unless unavoidable and unless they are replaced by perfectly identical goods" (Wissenburg 1998, 123).

But what if this environmental destruction is "unavoidable" in order to secure other goods which individuals prefer? The problem here is that, beyond bedrock liberal principles, all there is are individual preferences; environmental "concerns" are but one type of such preferences. If that is what people prefer, so be it; if not, then liberalism cannot override their choice:

> Now whether liberal democracy is incompatible with green concerns depends first and foremost on whether the rules of this political sphere allow green politics to be considered, designed, and implemented. If the rules do not allow this, green preferences are irrelevant because they simply cannot be translated into the appropriate rights.
>
> *(Wissenburg 1998, 10)*

The problem here is that a liberal society cannot (at least, in theory) encourage or discourage individual preferences: "Liberal democratic political systems have a prima facie duty not to interfere with the genesis of preferences" (Wissenburg 1998, 16). But what, then, generates preferences in individuals? As Robyn Eckersley points out in her criticism of Wissenburg, he assumes that individuals are free to make, and can make, rational choices about their preferences. All preferences then have equal value, and liberal polities should endeavor not to favor one or another. But the idea that social structures, or governments, or cultures, may themselves influence the preferences of individuals is not acknowledged (Eckersley 2004, 97). So if a political regime seeks greater resources in order to provide greater consumer choice—and a wider range of possible life choices—for its citizens, there is nothing in this that liberal theory can criticize. As Stephens concludes about this debate, "The criticism that a neutralist liberalism cannot *in practice* motivate green change or oppose runaway instrumentalization because of its refusal to examine preference formation mechanisms does not seem to me to have yet been adequately answered" (Stephens 2016, 64).

What this means is that democratic and liberal theories cannot readily, or perhaps even eventually, justify political regimes that would necessarily prohibit extractivism, especially if such ecosystem destruction serves the preferences of many individuals in a liberal state. Such preferences might be in favor of a highly industrialized and urbanized consumer society such as a few countries today have attained (e.g., in North America and Europe), and large political constituencies in others seek to attain (e.g., China, India). Political rhetoric endorsing growth and development are pervasive in many parts of the world, and constitute a key impediment to reining in the environmentally destructive forces of hypermodernization. Liberal societies do not have the conceptual resources—on either version of liberalism—to resist this, and in the neutrality-oriented version may even abet it.

The State as an Environmental Governance Institution

Perhaps it is the case that neither liberal democracy—or democratic theory and political liberalism separately—cannot provide strong rationales for ending extractivism and related environmental harms. Yet, there could be a different sort of argument for doing this—an argument not incompatible with liberal and/or democratic theories, though not necessarily mandated by them. This would be an argument from collective self-interest: the idea that, either for simple survival or for improved quality of life, it is imperative for states to adopt environmentally rational policies. Such policies would be conservationist and err on the side of caution and preservation concerning intact ecosystems. The argument has taken two forms. On the one hand, some have maintained that the state, because it is charged with solving problems that threaten or undermine collective well-being, must transform itself into an "ecostate," a state dedicated to the preservation of the ecological conditions for that would make such well-being possible. On the other hand, states may also be seen as acting, beyond simple security and survival, to further the welfare of the society, and the preservation of environmental goods is instrumental to that. Both of these arguments, however, must ultimately confront the problem that the state is also a mechanism for collective aggrandizement and enrichment, and to that extent, will be motivated to act to further some of the very policies that threaten ecosystems and their inhabitants.

Marius de Geus, in his article, "The Ecological Restructuring of the State," makes the first case for an "ecostate," on what might be called functionalist grounds (de Geus 1996). James Meadowcroft makes the second case, on welfarist grounds, in his article, "From Welfare State to Ecostate" (Meadowcroft 2005). De Geus emphasizes that the liberal democratic state is not to be surpassed or replaced by an ecostate, but simply "restructured" (de Geus 1996, 204). This means that an ecostate must be established in such a way that it "is maximally based on consent and . . . applies coercion minimally" (de Geus 1996, 206). Nevertheless, it must be established, in order to deal with environmental problems that are increasingly threatening to the well-being of society and possibly to the stability of the state itself. De Geus contrasts the ecostate with an ecoanarchist approach that seeks the abolition of the state, in order to establish a steady-state society. But de Geus maintains that the state should not be abolished, since it plays a crucial role in limiting and regulating market economies so that they do not overwhelm the ecological preconditions of their functioning. States should tolerate and indeed maintain market economies, but only under "strict ecological limiting conditions" (de Geus 1996, 201). These limiting conditions include several standard environmentalist principles that guide the state's (and the market's) restructuring: prevention (of environmental harms), a precautionary principle, legal liability for environmental harms, the "polluter pays" principle to cover the costs of "externalities," and a cautionary principle with regard to natural environments (this last being the most applicable for criticism of resource extractions) (de Geus 1996, 201–04).

The reason to restructure the state in these ways is that states are the best "problem-solvers" when it comes to such problems as environmental destruction. This is the case when: [1] the problems concern collective, not individual, goods; [2] the causal agents of the problems—in this case, industrial enterprises, largely—will not voluntarily reduce or eliminate the harms leading to the problem; [3] there is a competitive environment in which individual firms seek to reduce their culpability for the problem; and [4] there is a need for impartial expertise to solve the problem (de Geus 1996, 189–92). Only the state, on this view, can satisfy the conditions required to deal effectively with environmental problems.

While this view relies on the undeniable premise that the state *can* be an effective counterweight to the forces and agents of a market economy, it fails to demonstrate, or give any grounds for believing, that the state *will* fulfill this role. We can put aside for now the issue addressed earlier, that is, that the liberal and/or democratic state is especially ill-suited to establish limits to, or prohibitions on, degradation or destruction of ecosystems, for the reasons that such actions may serve popular wishes or initiatives for increased wealth or property ownership, or provide increased opportunities to realize a number of individual life plans. But even in terms of the state itself, it is not clear that states will necessarily adopt an interventionist or antagonistic stance to markets and economic agents when they perpetrate environmental harms. Part of the reason for this is that there is no acknowledgment of the fact that states have interests of their own—they are not simply neutral or impartial vehicles for solving problems or discharging functions in the general interest. States seek their own greater power and authority; they also seek revenue. Business and industry can contribute to these, as long as there is a cooperative, not antagonistic, relation between them and states. There are, in other words, material reasons why states may not undertake the regulatory function that de Geus argues is fundamental to them (for ecological problems, or more generally). States are often complicit in processes of capital accumulation or economic development, and are hardly in the best position to limit these processes when necessary for other reasons.

Of course, this has also seemed to be true in the past for other "social" problems, such as those of poverty or inequality. It is true that states *sometimes* take an adversarial role to business and industry, in order to secure improvements in social welfare. This is Meadowcroft's point in making the welfarist argument for the ecostate. If it could be done for other issues or causes, why not for the environment? This view relies on a comparison between previous social movements, and the concessions that they have gotten from liberal states, and the contemporary environmental movement, which seeks to force such states to act to prevent or end harms to ecosystems. That such environmental harms affect collective well-being as well is the working assumption of this argument: "the ecological state is predicated on a recognition that environmental systems are critical to long-term social welfare" (Meadowcroft 2005, 4).

But how valid is the comparison between earlier social movements (and the ways in which they motivated the creation of a welfare state) and the current environmental movement? This is the basis of the welfarist argument for an ecostate. Meadowcroft points to both similarities and differences, though he ultimately thinks the former outweigh the latter. Both welfare and ecological states involve the extension of state authority to new domains of policymaking. These extensions are based on claims of justice—whether social or environmental. Both welfare states and ecostates respond to market failures to secure certain goods for their citizens (social welfare, clean environments). Finally, the origins of these movements and their demands shaping new forms of the state lie in ongoing processes of economic development and innovation, and changes in the forms of capital accumulation that they entail (Meadowcroft 2005, 6–9).

At the same time, as Meadowcroft admits, the differences are marked. First, there are no social classes or groups (i.e., workers or the unemployed poor) associated with the movement to create the ecostate, as there was for the welfare state. Even more important, while a welfarist transformation of states was achieved in the midst of post-World War II economic growth, the ecostate is to be created in the absence of any such increases in private profits (other than those tied to financialization) or public revenues. While welfare states are clearly territorial, providing new benefits to a defined citizenry, ecostates must cooperate and coordinate environmental preservation across state boundaries. To the extent that both forms of the state intervene in markets, the type of intervention is different as well—taxation in the case of welfare, regulation in the case of the environment. Finally, while, perhaps as a result of its close ties to class movements and parties, welfare states resulted from increased democratic initiatives, environmental protection does not necessarily accord with majoritarian democracy.

All of this, as Meadowcroft admits, suggests that the welfare/ecology comparison may be strained. Despite the fact that social welfare has environmental preconditions, this does not necessarily mean that this is *enough* of a reason, given the various democratic, liberal, economic, and other arguments against it, to justify an ecostate. Meadowcroft's response is twofold. First, the transition to an ecostate will be prolonged, tentative, and always contestable. But it will presumably be necessary because of the press of events—the developing environmental crisis will create demands for action (Meadowcroft 2005, 18–21). Second, though there will be "headwinds" in the absence of a direct connection between economic growth or development and ecological regulation and restraint, there will nevertheless be gradual progress toward an ecostate as a result of the "*decoupling* of economic performance from *material* impacts, so that social welfare could rise even though resource inputs and waste levels fall" (Meadowcroft 2005, 11).

These points are well taken, but they do not rise to the level of a justification for an ecological state as the *primary* medium for addressing environmental crises, including extractivism. The reasons are these. First, there is a perceptual problem concerning the idea that social welfare or "progress" can be achieved without economic growth. This idea has been a mainstay of environmental economics for over a generation now, as in Herman Daly's idea that the meanings of economic growth and of development are distinct—the first being an increase in the quantity of goods and services produced, the second an increase in the efficiency with which goods and services of any quantity are produced (Daly 1996, 31). But that is a different matter from whether growth is *perceived* to be unnecessary for the more efficient satisfaction of human wants. The ubiquity of "growth" as a panacea endorsed by politicians in most countries across the ideological spectrum suggests that this difference will take a long time to be absorbed into popular consciousness. In the meantime, any state attempting to curtail economic growth will face sharp political consequences. Such an ecostate will be more than contingent on political exigencies; it will be seriously endangered, if it ever attempts to carry out real ecologically oriented changes that challenge economic orthodoxy.

This brings us to the second reason why the state cannot readily transform into an ecostate, a reason having to do with motivation. In the absence of a strong political movement challenging the ideology of economic growth, what other political constituency will be powerful enough to force a change? It is likely that, without a political movement opposing extractivism and advocating sovereignty for ecosystem peoples, whatever motivation that states and political elites may have to serve the requirements of ecological sustainability will be insufficient. In other words, states will follow, not lead, the transition—they cannot play an important role absent political movements that will force them to do so, as happened in the transformation of liberal states into welfare states.

Is an Ecological *Emergency* State Necessary?

The problems with thinking about the state in liberal-democratic terms, or in terms of its functional roles in relation to market economies or social movements, has led to two sorts of reactions. On the one hand, some have thought that a "stronger" state is necessary in order to deal with environmental problems. On the other hand, others have concluded that the state is greatly compromised in its ability or intention to deal with extractivism, ecocide, climate change, and other environmental problems, and therefore that other political agencies "below" or alongside states must play the crucial role. The concept of ecosovereignty is the result of the second kind of reaction, which this book exemplifies. But before we consider more concretely what sorts of institutions could make claims to ecosovereignty, the first reaction—sometimes regarded as "ecoauthoritarian"—needs examination. The general idea is that, in the face of an "ecological emergency," or imminent environmental collapse, perhaps caused by global warming, or by the depletion of resources through long-term extractivist practices, only a non-liberal, nondemocratic government would be able to use its power to make the hard decisions necessary to stave off catastrophe.

The question of whether democracy (including "liberal democracy") could be a suitable form of political theory and practice to deal with ecological emergencies was first raised most forcefully by William Ophuls, in his 1977 book, *Ecology and the Politics of Scarcity*. Despite much subsequent avoidance or dismissal of Ophuls' work, it remains among the most trenchant and perceptive studies of the problems of finding a suitable form of environmental governance, especially within the confines of modern, liberal democratic states. One of Ophuls' main points in the book was that, even if certain environmental conditions were necessary goods for the practice of democracy (not to say, the survival of human communities), this does not mean that democracies would recognize and valorize such conditions—especially if it meant giving up other goods cherished by their self-interested citizens. While survival conditions for societies are obviously fundamental, that does not mean that they will necessarily be foremost in citizens' interests or motivations.

Ophuls posed a question for liberal-democratic theorists who sought to use states to solve environmental problems: how much coercion was necessary in order to deal with the increasing scarcities resulting from extractivism (Ophuls 1977, 151)? He saw this, in fact, as an example of the "core problem of political philosophy"—when were limitations on persons' freedoms necessary, and how ought they be exercised? Coercion, as he pointed out, was not necessarily a bad thing, merely a means of achieving certain collective goals that required overriding individual self-interest. But in certain cases—of which ecological scarcity is the most extreme—a sufficient degree of coercion could sabotage other political goals. Self-government, the ostensible goal of democracy, was one of these goals.

What was the exact nature of this problem? It was the need for the citizenry of states to practice self-restraint, including the practice of virtues of temperance, modesty, and sufficiency, that seemed to require the exercise of coercive state power in order to override the selfish interests of democratic majorities. Ophuls believed that only a "steady-state society" could do what was required for ecological survival. Could such a society maintain a democratic state, that is, one dedicated to serving the wishes of the majorities of its citizens? Or would it be too absorbed in the temporary and immediate satisfactions of its people's short-term interests, while ignoring or downplaying the measures necessary for long-term survival? Because such societies were dedicated to satisfying either majoritarian or individualistic desires for greater wealth or consumption or leisure (depending on the version of democracy and/or liberalism that is institutionalized), states would be unable to restrain their populations without a substantial increase in their coercive power.

If this contradiction led to increasing scarcities of essential goods, there would be no way to manage the resulting problems without a much higher degree of coercion and abrogation of democratic rights and liberal entitlements. In other words, as Ophuls put it, "under conditions of ecological scarcity . . . democracy as we know it cannot conceivably survive" (Ophuls 1977, 152). Was there an alternative? Actually, there were two: but Ophuls, despite his own preferences, thought only one of these was probable without massive changes in existing societies. He himself believed that only a society consisting of "relatively small, intimate, locally autonomous, and self-governing communities rooted in the land" could ultimately achieve a modicum of ecological sustainability (Ophuls 1977, 241). But for "modern," complex societies, dedicated to the realization of individuals' interests without restraint, another option was more likely: only "centralization and expert control" could achieve the "steady state" necessary for long-term survival (Ophuls 1977, 162).

The reason for this dire conclusion is that *some* form of coercion would be required in order to rein in the self-destructive impulses of industrial society. Otherwise, nature itself would "coerce" human beings into accepting the biophysical limits within which life on earth would have to be lived. Aside from this, people needed either to learn self-restraint in order to live sustainably, or suffer "an iron regime that will compel our consent to living life with less" (Ophuls 1977, 156). How would such a regime operate? It would require a "special class of experts in charge of our survival and well-being—a 'priesthood of responsible technologists'" (Ophuls 1977, 159). They will manage the "highly-developed, complex technological civilization operating at or very near the ecological margin [of survival]"; only such management will forestall a collapse, given the demands of an overpopulated world, the complex technologies required to meet them, and the everpresent dangers of miscalculation implicit in the situation (Ophuls 1977, 160).

Lest Ophuls' view seems too bleak for the present moment (though how could it?), it is echoed by another quite recent treatment of "ecoauthoritarianism," that of Ross Mittiga. Mittiga argues that, while in "normal" times, governments earn their authority by providing their citizens with a number of rights and benefits, under "emergency" conditions, another type of legitimation of state authority is necessary (Mittiga 2022, 1001). This is what he calls "foundational legitimacy," and it "requires that citizens' essential safety needs are met" (Mittiga 2022, 999). In conditions of wartime, health emergencies, and now environmental or climate emergencies, governments can claim this foundational legitimacy if they are seen as taking necessary measures to deal with such situations. This is because "ensuring safety and security is always the *primary*—though . . . not the *exclusive*—end of political power" (Mittiga 2022, 1004).

Once having achieved this (foundational) legitimacy, maintaining it may require an "infusion of authoritarianism" (Mittiga 2022, 1005). Withholding (economic) benefits, suspending (political) rights, and other measures would be justified—and very likely perceived to be so by the populace—if dire enough conditions seemed to warrant them. Should we wish to avoid such a scenario, "we must do all we can to prevent emergencies from arising that can only be solved with such means" (Mittiga 2022, 999). Of course, the means themselves may still be inadequate to the task—or more to the point in this instance, too late. Some measures considered to deal with the "climate emergency" resulting from global warming would include mandated "lifestyle changes" (e.g., curbs on "meat-heavy diets"), censorship of "climate denialism," and the waiving of property rights to shut down "carbon-intensive production" by private companies (Mittiga 2022, 1007).

Will these be enough—are even authoritarian states equal to the task? Whether, as Ophuls writes, a technocratic priesthood armed with unlimited state power can stop the juggernaut of resource extractions that continues to power economic growth and "development" (from which only the few benefit) is doubtful. More probable might be the advent of a "green" industrialism powered by different, but just as intensive forms of extractivism and ecocide. In any case, the possibility of an ecological emergency state, whether or not it "succeeds," raises in a different way from the hopes for a liberal-democratic ecostate the limitations of environmental governance that is state-oriented.

There is no doubt that states are powerful mechanisms of collective action that could, in some possible worlds, solve coordination problems such as the environmental crisis of the twenty-first century. But the present moment is probably not one of those possible worlds, for the reasons just presented. For one thing, the "perfect storm" of the environmental crisis seems uniquely able to expose the weaknesses and contradictions of liberal democracies: the assumption of individual self-interest that sabotages collective decision-making, the majoritarian biases that devastate non-conforming marginal communities, and the formation of elites that accompany the growth of governments and are uniquely able to manipulate the processes of opinion and policy formation.

Furthermore, while in the past, state structures have sometimes been modified by political movements seeking social changes of various kinds, the problem of environmental destruction may be harder to solve in this way. This is because it strikes at the heart of state power, which is based on resource mobilization in service to state imperatives. How would a policy of *curbing* extractions, and thereby reducing the resources at the disposal of states, be something to which they would be committed? Beyond that, what reason is there to think that giving states even more power, in order to deal with an "emergency" situation, would actually serve this purpose? More likely—as in "Communist" China today, which has been engaged in a massive effort at capital accumulation—lip service would be given to a goal or ideal, while actions would be taken to achieve opposite ends.

But it is the case that states *are* sometimes susceptible to political pressure, *if* it is exercised by an awakened and militant body of citizens and citizen organizations. Here lies some hope for utilizing states to face the crisis before us. But it is entirely dependent on the ability of substate groups to hold their own in the face of economic coercion and political repression. This is where the ecosovereignty principle becomes important—as a justification for the power and legitimacy of such groups in their fight against extractivism and other ecocidal practices. The crucial basis for peoples' opposition to extractivism, and the social and environmental problems that follow from it, is their ability to control resources in the most concrete sense—land, water, energy—in order to provide the material basis for an alternative to extractivism.

Perhaps Ophuls is right that the real choice is between technocratic elites and small, autonomous communities dedicated to living more sustainably on the earth. But the former will crush the latter (as they are attempting to do in some places right now) unless there are political ideas—and resources organized on the basis of those ideas—that can mobilize support for resistance to these elites. So the question is: Given a principle of ecosovereignty, what institutions and forms of governance can be thought to embody it?

Common Property Institutions and Environmental Governance

This is where the importance of common property institutions (i.e., the commons) lies. While not usually thought to be "political" institutions, but rather alternatives to private property regimes, the commons today (in various iterations) is becoming increasingly important as a governance institution. This is not an entirely new idea, but it does herald a new understanding of "the political" as not necessarily synonymous with state power. The attempt to govern outside of states has, in modern times, usually been regarded as "anti-political" or apolitical. For instance, peasant communes, former slaves' communities, farmers' cooperatives, and indigenous tribes and peoples have often been regarded as little more than unorganized and spontaneous forms of collective life, in areas where states have not been established.

Historians today increasingly acknowledge that struggles to create such communities have existed throughout the modern era (Wolf 1969; Linebaugh 2014). Moreover, while traditionally regarded as movements seeking state power through the overthrow of existing regimes, they are increasingly understood as attempts to force the withdrawal of states from places in which other means of governance (i.e., common property regimes) were operative. What these have been or might be will be considered next. For the moment, such initiatives can be viewed as alternatives to state governance, rather than attempts to occupy or influence state regimes.

The recognition that states could not provide their own legitimation, which depended on some form of collective expression external to states' authority, has been a feature of modern political philosophy from Locke to Marx. While none of the best-known political philosophers, except for Marx, clearly rejected state authority, Locke, Rousseau, Hegel, and others clearly circumscribed it through making states subordinate to ideas that alone could legitimate their power. Whether it was Locke's right to private property, Rousseau's notion of the general will, or Hegel's concept of ethical life, state authority was justified only to the extent that it embodied one of these ideas. But that is not to say that these or other modern philosophers sought to legitimate *rival* forms of political power.

Only socialists such as Marx or Engels, on the one hand, or anarchists such as Bakunin or Kropotkin, on the other, clearly rejected the legitimacy of state power. But they too did not espouse any alternative to states as a form of *governance*. In this sense, the idea that commons institutions could circumscribe, and even provide a limited alternative to, state authority is a largely unanticipated idea. This can be seen even in contemporary anarchist thinkers who are otherwise sympathetic to environmentalism, such as Murray Bookchin.

Bookchin, in advocating his theory of "social ecology," has written about anarchism as being the expression of nonhierarchical relations between and among species in nature. The design of an "ecological society" finds inspiration in "preliterate societies," as well as from utopian socialist communities in the nineteenth century (Bookchin 1991, 318–20). Such a society, Bookchin claims, can operate without states or other hierarchical organizations, instead creating "libertarian institutions" based on direct democracy and face-to-face interactions (Bookchin 1991, 336–40).

The problem with this anarchist vision of communal societies is that it does not solve the "coordination problem"—how to coordinate between different individuals with distinct desires and needs. States and markets—and commons—are mechanisms for doing this. What does anarchism substitute for them? As André Gorz has written, in the absence of some institutional mechanisms, only social conformity enforced by an ideological unanimity (e.g., dogmatically held religious beliefs) can make such coordination possible (Gorz 1982, 108–12). Anarchism is valuable in suggesting that states are not an absolutely necessary feature of social life, and may be largely unsuited to constructing an ecologically sustainable human society. But it fails at proposing an alternative form of governance that can actually maintain (and defend) an ecologically equitable form of life.

A more recent source of thought about this is the theory of "bioregionalism," as developed by environmentalists in the last two generations. The idea of bioregionalism is that the natural context of human life is that of "bioregions"—areas united by common biophysical features (e.g., ecosystems, watersheds) (Sale 1985; see also Ribó 2012). This idea has had wide influence as a general characterization of human societies as socioecological systems (Crist 2019; Nine 2022). Its influence extends to this book; certainly much of what is recommended here requires acknowledging the importance of defining human societies in relation to the ecosystems—bioregions—that provide their natural foundations. There is little to quarrel with when bioregionalism is understood as a social theory.

But its political ramifications are less clear. Presumably, if human societies are based mainly on such bioregions, then states that span multiple regions are illegitimate, since they exist solely as a coercive structure imposed from without on particular bioregions. Yet, discussions of political, especially governance, arrangements for bioregions have not gotten very far. Kirkpatrick Sale, for instance, writes mainly of "communities" based on bioregions as the proper scale for social organization (Sale 1985, 62–64). Sale has much to say about the geographical scale that is optimal for bioregionally based communities (Sale 1985, 52–66). But the forms of political organization that are suitable for coordinating and administering such communities are left largely unspecified.

Principles of decentralization and territoriality are, to be sure, important parts of such a bioregional perspective (Sale 1985, 91–92). But beyond this, there is little specific to recommend, due to the laudable diversity of bioregions and the need for varied adaptations to their environments. At one point, Sale suggests that bioregional principles imply the desirability of "libertarian, noncoercive, open, and more-or-less democratic governance" (Sale 1985, 108). But he is quick to add that this is not a necessity, only a possibility. The main problem, though, is the lack of engagement with the actual politics of the current moment. How a transition occurs from the overcentralized, overindustrialized societies of the present to a world of bioregional "communities" is hard to imagine in the near term. What "conceptual tools" will help people to understand the goal and how to achieve it? A different approach is needed to theorize the best form of governance to ensure ecological equity (i.e., environmental justice) that will be compatible with such bioregional communities.

For this, we need to turn to recent theorizations of the commons as a distinctive form of social coordination and natural-human relations. This theory can be found especially in the work of the social scientist, Elinor Ostrom, among others (Ostrom 1990). Ostrom's work was developed in response to the earlier theorization of the commons by Garrett Hardin (Hardin 1968). Hardin's work eventuated in the enormously influential, but also seriously misleading, concept of the "tragedy of the commons," which from Ostrom's point of view was not concerned with the (real) commons.

Hardin's idea of the commons was that they constituted "open-access resources"—environmental goods or services that were open to any users. This fact meant that users, individually or collectively, could deplete the resource without any restraining forces being brought to bear to prevent it. Hardin's eventual conclusion was that only the privatization of such resources could protect them from such overuse. But from Ostrom's perspective, Hardin's view confuses open-access resources with commons *institutions*. The commons is a non-state, non-market means of managing natural goods and services so that they do not remain, or become, open-access resources. To the extent that a commons exists in relation to a particular good (e.g., lands, waters, forests, fisheries) that good is no longer open to all users (Ostrom 1990). Only when it is the case that neither the commons, the market, nor the state have taken responsibility for management of a natural good or service can a resource be said to be open-access. There are such situations, but they are rarer than Hardin, or other commentators who have adopted his view (e.g., Ophuls) have claimed. Even in the absence of markets and states, commons institutions often exist to limit and restrain "free riders" or other users who seek to monopolize, and perhaps deplete, goods and services.

So how are commons to be best understood, and how might they constitute the basis for environmental governance? Commons are informal, but structured, relations of communication and regulation of a given natural good by its users. These users are generally those who utilize the good as a resource for their livelihoods (and not for profit, as in extractivism). There are many historical and contemporary examples of such arrangements, ranging from village commons (usually of farmland) in Europe, Russia, and North America, to pasture lands in West Asia and elsewhere, to rural commons in farmed or forested areas of South Asia, to indigenous lands in many world regions. Political ecologists continue to do research to uncover historical and contemporary examples of working commons institutions that sustainably manage environmental goods and services for their (human and nonhuman) users.

Conceptually, commons institutions require the operation of several features in order to flourish, in the absence (or by exclusion of) markets and states.

First, natural goods must be understood as "community property resources," rather than as unowned goods, private property, or even "public" goods. This is because, as Vandana Shiva writes, "Neither the state nor the market is a successful system for allowing the people to sustain the use of a natural resource over a long period of time" (Shiva 2020, 251).

Second, there must be a means for users of natural goods to communicate with one another and reach agreements that will serve as the basis for sustainable use. As Ostrom writes, using the terminology of rational choice theory,

> Simply enabling subjects to engage in face-to-face communication between decision rounds enables them to approach socially optimal harvesting levels rather than severely overharvesting the commons. In the face-to-face discussions, participants tend to discuss what they all should do and build norms . . . to encourage conformance.
>
> *(Ostrom 2007, 15183)*

Another way of putting this point is to understand that, for a commons to operate effectively, it must be based on a community (of users).

This does mean that there are situations in which commons do not operate, *and* in which neither states nor markets hold sway. The most striking example of this is deep ocean fishing (as opposed to coastal fishing, which is often governed by commons arrangements). In this and other such cases, a commons needs to be created in which the users will often *be* states or corporations that have an interest in the resources. There are successful examples of this—for example, agreements about resource exploitation in Antarctica, or various global communications and transportation networks (postal, shipping, and air transport protocols) (Dolsak and Ostrom 2003). But most commons will continue to be small-scale, interactive arrangements between individual users who are members of an already existing community and who cooperate to sustainably manage the resource. Certainly, that has been the case historically.

The third feature of commons institutions is the most undeveloped, and often implicit in the arrangements, when not lacking entirely. This is the ability to preserve the natural goods which constitute a resource for members of the commons. Such ability is usually a function of the structure of the communities that are attached to the commons. For instance, villages, including local councils, may play this role. In South Asia, as an example, *panchayats* exist as councils to adjudicate disputes within a locality and to collectively manage local (natural) goods and services of interest to the villagers (Shiva 2020, 256–57). Similar institutions have existed in many other world regions. However, it is also important to note that many of these commons institutions have been destroyed or ceased to operate under external pressures of different sorts from markets, states, or both.

Such governance institutions, though informal, have played an essential role in preserving natural goods, and in sustainably using them to satisfy human needs. However, it is also the case that commons governance has been under attack since at least the seventeenth century in increasing parts of the world. A longer-term historical perspective might postulate a general antagonism between states and commons. But premodern states often existed either on a local scale, not unlike commons themselves, or on an imperial scale that not only transcended but also ignored the local scale of the commons. It is only with the modern state—and even more with the development of the capitalist market—that commons have been affected by sustained assaults by these other institutions. Marx, in his discussion of the enclosure movement in Britain in the seventeenth and eighteenth centuries, named these assaults "primitive accumulation," in the sense of an original accumulation of capital (i.e., land and labor) (Marx 1992). But in the past two centuries, attempts to enclose the commons have continued. States and corporations today wield both modern weaponry, legal mechanisms, and economic power to eliminate commons as an impediment to the accumulation of more capital.

This is the *real* tragedy of the commons. In many cases, commons have been able to function for long periods of time because there were no external pressures to introduce private or public property arrangements, from which others could profit. Many human societies largely existed outside of market or state relations, with peoples' livelihoods being dependent, not on such institutions, but on commons management of ecosystems. That is increasingly no longer the case. The tragedy is that so many common property institutions, on which the livelihood of many peoples continues to depend, are increasingly under withering assault by business enterprises engaged in continual resource extractions for profit, aided by states that facilitate (and themselves benefit from) such extractions. It seems to be the case that unique features of commons institutions, their informal and local character, may be their undoing in a world where they are under constant attack.

The story of this assault is one of land grabs, enclosures, and "roving bandits" stealing natural goods—and also, in some cases, of exterminations and extinctions (Winchester 2021; Shrubsole 2019; Olson 2000). But there is also a history of resistance to such attacks. This history is one that involves both spontaneous rebellions (e.g., the Chipko movement in North India), and large-scale social movements (e.g., the Zapatistas in the Yucatan), as well as legal efforts to protect ecosystems and the commons that manage them (Guha 2000; Shiva 2020, 280–309).

Yet, what is missing is any systematic attempt to formulate principles that can provide a long-term defense of common property and an effective opposition to efforts to plunder natural systems for profit-making goods and services. This is the purpose of the principle of ecosovereignty. It, in turn, needs embodiment in governance institutions with responsibilities to preserve natural systems for their (local) users and inhabitants. What kind of governance institutions? Here is a partial list:

small-scale states (e.g., ones limited to one ecosystem)
states (or provinces) within federal systems
local governments
community land trusts
village (or rural) commons
indigenous governments
(urban) neighborhood councils
resource users' collectives (e.g., farmers' or fishers' collectives)

All of these bodies are instances of common property institutions; they are also, *potentially*, environmental governance institutions. To the extent that they claim a right to ecosovereignty over the natural goods that they are capable of managing and sustaining, they are entitled to exercise, not only some authority over these goods but *ultimate* authority over them. While there are important differences between these types of commons, it is the similarities that should be notable as well: their ties to particular ecosystems and their ability to manage the resources for the livelihood of local users—rather than extracting them for their profit-making potential.

Environmental governance, then, is, above all, a feature of highly local management for particular (livelihood-oriented) usage, legitimated by a principle that accords ultimate authority to those who will exercise it in accord with values of sustainability (of the ecosystem) and sufficiency (of human use).

Ecosovereignty and the Natural Constitution

To recall, the principle (of ecosovereignty) is the following:

> Peoples living in distinct territories have ultimate authority over matters affecting environmental conservation in order to sustainably maintain environmental goods and services in those territories.

This principle is designed to provide legitimacy to common property institutions that seek to defend natural environments from resource extractions and to thereby achieve a measure of ecological equity. By providing a rationale for commons to assert their *political* rights, the ecosovereignty principle justifies efforts on the part of peoples—indigenous, place-based, subsistence-oriented, land-poor—who inhabit particular ecosystems to obtain the autonomy necessary to sustain and defend them.

The main purpose of the ecosovereignty principle is to provide a way for contemporary social movements of indigenous and working peoples to stop the extractivism that is robbing them of the lands and territories they need in order for their communities to survive. But the significance of the principle is greater than this, in part because of the effects of the extractivism that it is designed to halt. Such resource extractions are not only the principal cause of environmental injustices, in depriving peoples of the ability to sustain themselves without subsumption in industrial societies. They are also the main origin point of the fossil-fuel economy that is driving global warming, as well as one of the principal causes of habitat loss that is producing species extinctions and ecocide more generally. While extractivism seems to be a specific problem related to resource depletion, its ramifications are world-destroying. Halting, or even slowing, the process of extractions is essential for several reasons having to do with environmental justice, ecosystem health, and the survival conditions for many living species.

But it would be disingenuous not to consider the possibility that a philosophical principle alone, even one that can justify a political movement to achieve equity and justice, cannot achieve such goals. Put differently, the commons institutions that can give voice to such a principle may not be able to achieve a lasting solution by themselves. Arrayed against the combined might of state power and market forces, commons have often not succeeded in defending their lands and territories against the encroachment of extractivist forces.

States, inasmuch as they are susceptible to modification or limitation in their support for extractivism, must therefore be enlisted or cajoled into according sovereignty to the ecosystem peoples deserving of it. To do this in any thoroughgoing way becomes a struggle for constitutional revision and recognition. The first stage in such a struggle against extractivism is to mobilize peoples to claim sovereignty rights and to defend their environments on the basis of that claim. But the second stage is one of achieving constitutional recognition of the "rights of nature" and of those peoples who have an intrinsic and sustainable relationship to nature. It is, in other words, to establish a "natural constitution" that recognizes a principle of ecosovereignty and the corresponding territorial rights and common property regimes that necessarily accompany that right.

The idea of a natural constitution is not a new one. It has been expressed in various contexts and has even been instantiated in a few instances, such as constitutional provisions protecting nonhuman species, ecosystems, and indigenous peoples, in such places as India, Canada, and some South and Central American states. A natural constitution need not be a new one entirely; it can involve amendment of existing constitutions. Earlier proposals for such amendments generally concerned such things as the protection of certain natural resources, or the biological diversity of an environment or region (Norton 2005, 499, 502).

As discussed in Chapter 2, the recognition of indigenous rights has also become a part of some state constitutions. But to recognize a claim to ecosovereignty, or even a more limited claim of territorial rights or common property arrangements, would involve constitutional ratification of changes in *basic* constitutional provisions having to do with who holds ultimate authority over what matters.

While this idea of a natural constitution would therefore seem to be a more difficult achievement, it would have both practical and theoretical advantages over sovereignty claims that did not achieve (constitutional) recognition, were it to be accomplished. Politically, it would bind those who have valid claims to (eco)sovereignty with all citizens, who would ostensibly recognize the validity of the claim-right. It would also provide a powerful rationale for the constitutional state to support and defend common property regimes, at least in certain areas or regions accorded rights to ecosovereignty. While commons, as discussed earlier, are at a disadvantage in ecological distribution conflicts when both state power and markets are deployed against them, if states were to recognize common property regimes as legitimate, this would be a very significant alteration in the power dynamics of such conflicts. In the end, an ecosovereignty principle whose legitimacy was always subject to a particular balance of political power would be more fragile and potentially temporary than one that had achieved the status of a constitutional principle for the states to which it applied.

Theoretically, aspects of such a natural constitution—one that recognized an ecosovereignty principle—would have two advantages over sovereignty rights that did not receive constitutionalization. First, it would be the recognition, as Bryan Norton writes, that identifying "features of the physical landscape for protection *constitutes* a community as a place, a place considered 'home' by its inhabitants whose activities are embedded in a particular physical system, or habitat" (Norton 2005, 501). Whether or not the places recognized as such equal the whole of the state (as in small states comprising one ecosystem) or a part of the state, as in ecoregions within larger states, would make some difference, but not as a matter of principle. In the latter case, constitutional provisions would bind the particular regions to the larger state, inasmuch as the region obtained the protections of ecosovereignty for its environment, and the citizens of the larger state recognized the value of maintaining that ecosystem (rather than plundering it for resources). Second, making constitutional provision for the rights of ecosystems and their peoples would be equivalent to a "bequest for future generations" in which the value of certain (environmental) goods are recognized in perpetuity (Norton 2005, 499). This, in other words, would constitute a further form of institutionalization, beyond the ad hoc resolutions of EDCs.

For these reasons, the institutionalization of ecosovereignty as a constitutional right represents a further step in the recognition and protection of natural environments and the peoples who depend upon them. It also means that the idea of society as a *natural*, or socioecological, system has been accepted and accorded important value by the state and its citizens as a whole. Here, then, is a way in which ecosovereignty can play a role in substantially altering the balance of power in political systems without having to take on the burden of changing everything. Constitutional recognition would be the endpoint of a successful struggle for environmental justice, and the achievement of a hopefully enduring system of environmental governance.

References

Beitz, Charles. 1999. *Political Theory and International Relations.* Revised ed. Princeton, NJ: Princeton University Press.

Bookchin, Murray. 1991. *The Ecology of Freedom: The Emergence and Dissolution of Hierarchy.* Montreal: Black Rose Books.

Caney, Simon. 2016. "Cosmopolitanism and the Environment." In *The Oxford Handbook of Environmental Political Theory*, edited by Teena Gabrielson, Cheryl Hall, John M. Meyer, and David Schlosberg. Oxford: Oxford University Press.

Crist, Eileen. 2019. *Abundant Earth: Toward an Ecological Civilization.* Chicago, IL: University of Chicago Press.

Dahbour, Omar. 2019. "Justice, Social Not Global." *Radical Philosophy Review* 22 (1): 31–58.

Daly, Herman. 1996. *Beyond Growth: The Economics of Sustainable Development.* Boston, MA: Beacon Press.

De Geus, Marius. 1996. "The Ecological Restructuring of the State." In *Democracy and Green Political Thought: Sustainability, Rights and Citizenship*, edited by Brian Doherty and Marius de Geus. London: Routledge.

Dolsak, Nives, and Elinor Ostrom. 2003. "The Challenge of the Commons." In *The Commons in the New Millenium: Challenges and Adaptations*, edited by Nives Dolsak and Elinor Ostrom. Cambridge: Massachusetts Institute of Technology Press. Ecocidelaw.com.

Eckersley, Robyn. 1996. "Greening Liberal Democracy: The Rights Discourse Revisited." In *Democracy and Green Political Thought: Sustainability, Rights and Citizenship*, edited by Brian Doherty and Marius de Geus. London: Routledge.

Eckersley, Robyn. 2004. *The Green State: Rethinking Democracy and Sovereignty.* Cambridge: Massachusetts Institute of Technology Press.

Gardner, Steven. 2013. *A Perfect Moral Storm: The Ethical Tragedy of Climate Change.* Oxford: Oxford University Press.

Goodin, Robert. 1992. *Green Political Theory.* Cambridge: Polity Press.

Gorz, André. 1982. *Farewell to the Working Class: An Essay on Post-Industrial Socialism.* Boston, MA: South End Press.

Guha, Ramachandra. 2000. "This Fissured Land: An Ecological History of India." In *The Use and Abuse of Nature*, edited by Madhav Gadgil and Ramachandra Guha. Delhi: Oxford University Press.

Hardin, Garrett. 1968. "The Tragedy of the Commons: The Population Problem Has No Technical Solution; It Requires a Fundamental Extension in Morality." *Science* 162 (3859): 1243–48.

Jamieson, Dale. 2014. *Reason in Dark Times: Why the Struggle Against Climate Change Failed—And What It Means for Our Future.* Oxford: Oxford University Press.

Kolers, Avery. 2012. "Justice, Territory, and Natural Resources." *Political Studies* 60: 269–86.

Linebaugh, Peter. 2014. *Stop, Thief!: The Commons, Enclosures, and Resistance.* Oakland, CA: PM Press.
Marx, Karl. 1992. *Capital: A Critique of Political Economy*, Vol. 1, translated by Ben Fowkes. London: Penguin Books.
Meadowcroft, James. 2005. "From Welfare State to Ecostate." In *The State and Global Ecological Crisis*, edited by John Barry and Robyn Eckersley. Cambridge, MA: Massachusetts Institute of Technology Press.
Mill, John Stuart. 2004. "Of the Stationary State." In *Principles of Political Economy, with Some of Their Applications to Social Philosophy*, abridged ed. Indianapolis, IN: Hackett.
Mittiga, Ross. 2022. "Political Legitimacy, Authoritarianism, and Climate Change." *American Political Science Review* 116 (3): 998–1011.
Nine, Cara. 2022. *Sharing Territories: Overlapping Self-Determination and Resource Rights.* Oxford: Oxford University Press.
Norton, Bryan. 2005. *Sustainability: A Philosophy of Adaptive Ecosystem Management.* Chicago, IL: University of Chicago Press.
Olson, Mancur. 2000. *Power and Prosperity: Outgrowing Communist and Capitalist Dictatorships.* New York: Basic Books.
Ophuls, William. 1977. *Ecology and the Politics of Scarcity: Prologue to a Political Theory of the Steady State.* San Francisco, CA: W. H. Freeman.
Ostrom, Elinor. 1990. *Governing the Commons: The Evolution of Institutions for Collective Action.* Cambridge: Cambridge University Press.
Ostrom, Elinor. 2007. "A Diagnostic Approach for Going Beyond Panaceas." *Proceedings of the National Academy of Sciences* 104 (39): 15181–87.
Rawls, John. 1993. *Political Liberalism.* New York: Columbia University Press.
Rawls, John. 2001. *Justice as Fairness: A Restatement.* Cambridge, MA: Belknap Press.
Ribó, Ignasi. 2012. *Habitat: The Ecopolitical Nation.* London: Mycelia Books.
Sale, Kirkpatrick. 1985. *Dwellers in the Land: The Bioregional Vision.* San Francisco: Sierra Club Books.
Shiva, Vandana. 2020. *Reclaiming the Commons: Biodiversity, Indigenous Knowledge, and the Rights of Mother Earth.* Santa Fe, NM: Synergetic Press.
Shrubsole, Guy. 2019. *Who Owns England?: How We Lost Our Green and Pleasant Land and How to Take It Back.* London: Williams Collins.
Stephens, Piers. 2016. "Environmental Political Theory and the Liberal Tradition." In *The Oxford Handbook of Environmental Political Theory*, edited by Teena Gabrielson, Cheryl Hall, John M. Meyer, and David Schlosberg. Oxford: Oxford University Press.
Swart, Lydia, and Estelle Perry, eds. 2007. *Global Environmental Governance: Perspectives on the Current Debate.* New York: Center for UN Reform Education.
Winchester, Simon. 2021. *Land: How the Hunger for Ownership Shaped the Modern World.* New York: Harper.
Wissenburg, Marcel. 1998. *Green Liberalism: The Free and the Green Society.* London: UCL Press.
Wolf, Eric. 1969. *Peasant Wars of the Twentieth Century.* New York: Harper.
Woods, Keri. 2016. "Environmental Human Rights." In *The Oxford Handbook of Environmental Political Theory*, edited by Teena Gabrielson, Cheryl Hall, John M. Meyer, and David Schlosberg. Oxford: Oxford University Press.

CONCLUSION

Ecosovereignty, Collapse, and Transition

Ecosovereignty is a principle designed to justify political actions against extractivism and the untrammeled exploitation of nature. Such exploitation is rapidly leading much of the world's societies to a worst-case scenario of ecosystem loss, resource depletion, and societal collapse. There is a suite of other principles, concepts, and laws that have been developed to forestall such an eventuality. But without a politically potent force exerted on their behalf, such principles—ecological equity, ecological sustainability, food sovereignty, indigenous rights, resilience, sufficiency—will have little chance of realization. Philosophy can only provide the justification for, and legitimacy of, such a force. Yet, such efforts may motivate others to act in accordance with ideas that can provide a way forward that avoids the worst-case scenario just mentioned.

The principle of ecosovereignty has already been defined and justified; but some summary comments about it may be useful. Then, a skeptical doubt and a hopeful possibility will both be entertained by way of conclusion. The skeptical doubt suggests that the moment has already past for solutions that can avoid the large-scale collapse of much of modern society, at least in areas of the world suffering maximum environmental stress. In the face of such collapse, what can legitimating small-scale, ecoregional communities do to prevent the spread of disastrous conditions that will overwhelm societies attempting to limit the damage? The hopeful possibility to counterpose to this gloomy scenario is one of transition away from industrial capitalism, and to a new, more sustainable form of human-natural relations. What can ecosovereignty contribute to making this possibility a reality?

DOI: 10.4324/9781003568155-14

In defining ecosovereignty, the attempt was made to specify the principle across three dimensions, those of jurisdiction, membership, and authority. The type of jurisdiction that the principle mandates is one of regions, defined in terms of ecological characteristics—that is, as coextensive with an ecosystem. Ecosystems, of course, are malleable concepts, as with many scientific terms. The definition of ecosystems here will tend toward the maximal extent, in order to ensure as much (political) autonomy as possible for those within the jurisdictions. But the key point is that the jurisdiction is defined, in most cases, as a substate (eco) region. In some cases, such regions may be coextensive with (small) states; perhaps this is the easiest "spatial fix." But most contemporary states exist at a scale that contains more than one ecosystem, even broadly defined; for these, the jurisdictional authority legitimated by the ecosovereignty principle will be a smaller, substate region.

Membership in such regions will be defined relationally. But in many concepts of political membership, while they may also be relational, the relations are those between members of social groups, ethnic or racial categories, or historically associated peoples. Ecosovereignty legitimates membership that is defined in relation to natural environments, above all to land. While the general term for such membership is that of ecosystem peoples, the best-known example of such membership relations is indigenous peoples. But even here, as discussed in Chapter 2, many concepts of indigeneity rely on anthropological understandings of identity relations that are quite different from membership based on identification with lands or ecosystems. It is this latter form of relations that best defines the membership that is justified by ecosovereignty.

Finally, ecosovereignty defines a form of authority that is functional, rather than total or absolute. Since the purpose of the principle is to ensure ecological sustainability, other matters sometimes thought to be a part of sovereign authority can be put to the side. In such areas of concern as social welfare, public education, transport and communication, and diplomatic relations, ecosovereignty need not apply, and ecosystem peoples would not challenge the sovereignty of states in these matters, as long as the states themselves recognize limits to their own sovereignty when it comes to ecosystem maintenance. Ecosovereignty, then, is a claim of authority only for this purpose, by peoples with a particular concern to maintain such systems, and in regions clearly defined by the limits of such ecosystems.

Lastly, it should be reiterated that this aspect of ecosovereignty also establishes a limit condition on the exercise of authority. If such authority is used for purposes other than ecological sustainability, then sovereign claims are forfeited. In this sense, the principle does not accord with claims for *absolute* authority over natural resources, territory, or land, by peoples living there. It is only to the extent that authority is exercised to realize sustainability that it is legitimate.

Justifying the ecosovereignty principle involves three distinct claims: [1] it is the best current interpretation of the sovereignty principle generally; [2] it is necessary to ensure territorial rights for peoples in the face of rival claims; and [3] it best realizes equitable distributions of natural goods and the independence of common property institutions that themselves are necessary for realizing ecological equity. Before turning to the larger questions about the principle, these justifications will be briefly reviewed.

Ecosovereignty adopts the core meaning of sovereignty as ultimate authority over peoples and territories in relation to other sovereign entities—"external" sovereignty—and applies it to matters *internal* to states. In particular, it focuses attention on the relation between industrial capitalist economies and the sacrifice zones of resource extraction that supply those economies with their material prerequisites. The principle results from regarding this relation as one of "internal colonialism," a relation that legitimates claims of self-determination and sovereignty, similar to the anticolonial struggles of the twentieth century that legitimated sovereignty claims on the part of colonial territories. But in the twenty-first century, colonialism has become "internal" to large states with developed or developing industrial economies, and it concerns the exploitation of natural environments by extractivism just as much (or more) as it concerns the exploitation of labor. This is to say that ecosovereignty is to be understood as the contemporary expression of self-determination rights on the part of peoples seeking to maintain and protect ecosystems upon which they depend from exploitation by extractive industries.

As discussed in Chapters 1 and 2, such peoples include food producers and indigenous groups, as legitimate claimants of ecosovereignty. But it is important to note that other peoples connected to particular lands and territories would also be included. Important cases of this are peoples that seek autonomy within states, in order to pursue more sustainable goals for their region. Invoking ecosovereignty would, in a sense, become a way of "federalizing" the state, since regions would seek substantially more autonomy, at least in regard to environmental affairs, than they would have had hitherto. Claiming full state sovereignty in such instances would only be justified as a last resort, given the continued intransigence of the larger state to grant regional autonomy (Dahbour 2013, 187–93). The most egregious examples of such cases occur when the larger state is engaged in resource extractions that favor other regions (or even other states or societies). In such instances, the concept of "discriminatory redistribution," which justifies autonomy or even secession, would be relevant (Buchanan 1991). It is also important to remember that such cases are only justified as means of achieving ecological sustainability (Ribo 2012). Other arguments for federalism, autonomy, or independence based on purported rights to develop a region's resources would not apply (Nairn 1981, 126–95).

Beyond the meaning of sovereignty that the ecosovereignty principle embodies, it is also the case that it is justified as a means of validating territorial rights on the part of ecosystem peoples in the face of claims by purportedly "sovereign territorial states." Since states, in the absence of challenge, will assume absolute sovereignty rights over their territories, substate peoples must claim at least "functional" sovereignty over the natural environments with which they are connected in order to withstand extractive practices. While, as food producers, indigenous groups, and others have affirmed, an ecosovereignty claim need not go to the level of total independence, it must include territorial rights, including ultimate control over land and water rights and usage of resources contained in them. Otherwise, the core meaning of territorial rights themselves—the right to maintain and use natural resources for local livelihoods within a circumscribed territory—will risk violation and exploitation by other interests and powers, including states claiming sovereignty over them in order *encourage* resource extractions for profit.

In a sense, the ecosovereignty principle *politicizes* territorial rights, moving it definitively away from its possible conflation with property ownership, and legitimating territoriality on the basis of sovereignty for those peoples committed to sustainably managing territories. While, as just indicated, it is the case that food producers and indigenous peoples can make ecosovereignty claims to justify their territorial rights, they are not the only ones able to do so. Just as sovereignty claims by substate regions may, under certain conditions, be legitimate, so peoples or groups clearly asserting rights to preserve and maintain territories ecologically could make an ecosovereignty claim on the basis of their territorial rights.

One example of this is the recent upsurge in territorial rights movements in Europe, sometimes organized under the term, "earth uprisings" (*soulèvements de la terre*). These movements, especially strong in France, though not only there, organize on the basis of "the need to reclaim our material conditions of subsistence and to defend the terrestrial commons on which our existence and that of other forms of life depend: water, earth, air, wildlife [Sophie Gosselin]" (Dann 2023, 7). Such movements claim they are engaged in the (*self-)*defense of nature in specific territories, and invoke the idea of the commons for their own self-understanding. As a movement, then, the idea of territorial rights takes root in a combination of local peoples seeking to defend against extractivism and development, with others in neighboring cities and towns seeking to maintain nearby green spaces and water reserves. In this sense, they recall other movements that combine local, sometimes indigenous, groups with elements of urban populations committed to environmentalist projects (e.g., the Zapatistas in Mexico).

In addition to expressing the contemporary meanings of sovereignty and territoriality, ecosovereignty claims are necessary in order to legitimate governance bodies, such as common property institutions of different kinds, that can realize the equitable resolutions of conflicts over ecosystem goods and services. Other institutions, principally states, will tend to favor extractive enterprises that seek to use such natural resources for economic growth and industrial development. This is the result of the complicity of territorial states and corporations in the benefits they envisage from such extractivism. Only according legitimacy to governance institutions rooted in ecosystems themselves can redress the imbalances of political power that bias conflict resolutions over the distribution of resources toward extractive agencies. Examples given earlier include existing rural or village commons, community land trusts, and indigenous governing bodies. Ecosovereignty is the principle legitimating the claims of such common property and other governance institutions that seek to achieve true ecological equity and environmental justice in the distribution of resources.

It is crucial to realize here that the purpose of such governance institutions needs to be carefully circumscribed as the achievement of ecological equity in the resolution of distributive conflicts over natural resources. Equity, as understood here, means the prioritizing of attachments to, and needs for, ecosystem goods and services by peoples living in proximity to those ecosystems. Departing from stakeholder theory, contract theory, rational choice theory, and other paradigms of distributive justice used to determine *economic* equity, environmental justice in this meaning asserts the priority of communal relations *within* socioecological systems that are committed to long-term sustainability of natural goods and services. This is the purpose for which governance bodies are established that warrants ecosovereignty claims by various groups and peoples.

This, in brief, is the conception of, and rationale for, the principle of ecosovereignty. Yet, despite all this, it may still be hard to impossible to achieve ecological sustainability and equity, due to the mounting pressures from runaway anthropogenic environmental change. At least this is the worry of some who may be skeptical of the kind of solution offered here. There are two aspects to this worry: [1] the loss of natural habitats that can sustain a rich biodiversity, and [2] the multiple effects of anthropogenic "global heating" on lands and waters capable of supporting life.

There is no doubt that there has been and is continuing to be tremendous biodiversity loss, affecting multiple species and environments, up to and including extinction events. While extractivism is a huge contributor to this loss, and has been the focus of the discussion in this book, leading to advocacy of ecosovereignty, this is not the whole of the problem. Beyond extractivism, there is also the inadvertent loss of habitat—living space—for multiple species, including the degradation of ecosystems that are unable to support the biodiversity formerly present in them. The drivers of this loss are twofold. On the one hand, the tremendous expansion of human habitation across landscapes, due partly to simple demographic growth, but also to suburban and other land-intensive settlement patterns, is one. For instance, virtually all wild animal species in the most densely populated world regions (Europe, China) are approaching extinction. The same is true of marine species in prime ocean fishing grounds, as well as native plant species in many world regions. Not only is this the result of predatory human practices. It is also caused by the increasing loss of natural habitat to human habitation, as well as the disruption of natural corridors for animal migration and movement. All this (and more) is radically reducing nonhuman populations (Crist 2019, 154–60).

On the other hand, the development of built infrastructure to support urban-industrial societies has also taken a tremendous toll. Such infrastructure—roads, housing, energy production and transmission systems, estimated at four thousand *tons* per individual human being—has left little habitat for other species undisturbed (Purdy 2019, 77–84). Both problems originate in the drastic overpopulation that is now rapidly pushing the earth beyond its carrying capacity. Three dimensions of this problem are susceptible to amelioration, the last of which promises to be the distinctive contribution of ecosovereignty to halting the biodiversity and habitat loss affecting so many living species.

First, there are ways in which human overpopulation can be slowed, and eventually reversed. The problem is whether these ways will be adopted extensively by those political regimes that could make a difference. Evidence suggests that culture and education may make the most difference in affecting human population growth. Societies that have adopted small human families as cultural norms (China, Iran, and Thailand) have had success in stabilizing their populations (Cafaro and Crist 2012). But some of these societies have backtracked, once their ruling regimes focused on what were perceived to be the costs in geopolitical power that resulted from stable or declining populations. In addition, women's education specifically has a high correlation with stable populations. But the extent of commitment to providing women's education varies widely. While human fertility rates are declining, the increase in raw numbers is so large that the world's population is still expected to grow for at least two to three more decades.

Aside from the underlying problem of human overpopulation, there is a second possibility for reducing biodiversity loss in the recent attempts at habitat and species conservation and regeneration. Though it is undoubtedly the case that there has been widespread loss of wild plant and animal populations, it is possible to exaggerate how far this process has gone. There are still substantial wild populations, though often in remote places which lack interest for humans because of their remoteness, barrenness, or adverse climates (MacKinnon 2013, 142–46). More to the point, it is still possible to engage in large-scale conservation projects that would preserve, and even better regenerate, much of the biodiversity presently still existent (Wilson 2016, 175–76). Whether this will be done, however, is the question. It is highly likely that some of the species most prized by naturalists and others (e.g., polar bears) will not live out this century.

One aspect of this problem is whether the increasing loss of biodiversity will make it harder for humans to maintain their livelihoods on the basis of intact ecosystems. The key to answering this question lies, not in being able to limit human habitation at all, but in limiting the *effects* of such habitation by changing its character. This will be possible if the built or constructed infrastructure needed for human habitation is itself drastically reduced. Infrastructure is the core of the problem, then, not humans as such. This does not mean that proximity to human beings does not have negative effects on *some* wild species, especially large predators, delicate marine environments, and other fragile or complex ecosystems. Conservation biologists still see a need for wildlife preserves devoid of human presence; E. O. Wilson's "half-earth" proposal being the best-known example of this (Wilson 2016).

But it is also the case that many nonhuman species can live in close proximity to humans *if* human beings live in certain ways, and not others. It would be a serious error to adopt the view of ecomodernists that there needs to be even more construction of urban-industrial infrastructure, so that human beings can live separately from other species, who would survive untouched by the effects of human habitation. Where would the natural materials from which such infrastructure is constructed be found? How would they be procured? The answer is not hard to find: such an approach would require even more extractivism than is occurring at present, with increased exploitation of natural environments for their resources.

It would be one effect of widespread adoption of the ecosovereignty principle that greater populations would live in environments that do not require continual increases in resource extractions. Not only fewer overbuilt cities but also fewer roads, pipelines, industrial farms, mines, and forestry operations would mean more environments in which the human presence is tolerable for a wider range of other life. As discussed in Chapter 1, it is not human habitation per se, or human agriculture, that disturbs natural environments. It is *industrial* agriculture, mining, and other forms of resource extraction that decimate them. Sustainable agriculture, as well as forestry, aquaculture, and other forms of livelihood drawn from agrarian or other semi-cultivated environments can coexist with a high degree of biodiversity. Ecosovereignty then can be seen to be instrumental in ensuring that a "natural" infrastructure remains in place for humans to use and preserve—and that further construction of artificial infrastructure becomes superfluous to the needs of increasing numbers of peoples living more sustainably (Battistoni 2025).

Though biodiversity loss seems a massive problem, it is exacerbated—as so many other things will be—by the effects of anthropogenic climate change. The question here is whether the ecosystem peoples and ecoregions that are the beneficiaries of the ecosovereignty principle will be unable to maintain the natural systems upon which they depend in the face of global warming. Obviously, the first thing to say is that it depends on the geographical locations of the ecosystems. Certainly, in the tropical zones, there will be massive effects as global warming increases. Water loss and soil degradation are two major effects, as will be general biomass loss due to excessive heat. The effects of these processes will, in turn, be that large numbers of people will seek to migrate to cooler and still fertile areas in the temporate zones (i.e., in the North) (Vince 2022, 106). The timing and extent of these changes is unknowable, but probably occurring within the next three to four decades.

These are unprecedented and extraordinary geographic and demographic changes, and every socioecological system will struggle to cope with them. Can the attempt to maintain and protect ecosystems survive such wrenching transformations? Two problems in particular must be faced: what will people eat? Where will people live? The problem of feeding extra billions, especially if they are on the move, will be immense. But it is likely that the food system resulting from the industrialization of agriculture will be unable to cope, for a number of reasons, not least soil degradation, as well as water depletion (Pimentel and Burgess 2013; Gies 2022, 21–23). The waste of acreage for growing animal feed, for pasturage, and for biofuels, will have to be curtailed. Agroecological methods hold promise because of their intensive, diverse, and plant-based nature; certainly this will be one beneficial effect of protecting ecosystem peoples committed to this form of agriculture. Conservation of water and preservation of healthy soils using traditional methods will also aid food production. There are no guarantees one way or another; but it is a gamble worth taking, given the known deficiencies of industrial agriculture.

One approach to housing massive migrant populations will be to expand urban areas (in the North) to accommodate them (Vince 2022, 126–27). But this seems unlikely to be an effective solution for a couple of reasons. First, such a solution would require large government investments for urban development, something that has occurred only in China in recent decades, and is now ended even there. It is unlikely that states that have failed to take the measures necessary to prevent climate change will now take the measures necessary to manage its devastating effects (Gardner 2013; Jamieson 2014). More likely, without an alternative, people will settle in urban slums without services or employment, and subject to conditions of widespread exploitation, violence, malnutrition, and disease (Davis 2005). The alternative that is suggested by the ecosovereignty principle would be absorption of new populations in labor-intensive farming communities oriented toward local self-sufficiency.

This scenario leads to the consideration of whether the ecosovereignty principle could be an impetus to more far-reaching social changes that would herald the transition to a new and more sustainable world for both humans and nonhumans. Certainly, much more than a political principle asserting the rights of certain peoples to self-determination and sufficiency will be necessary. As Jonathan

Crary has recently written, "Any possible path to a survivable planet will be far more wrenching than most recognize or will openly admit" (Crary 2022, 4). The inability of "market mechanisms" and state agencies to ameliorate even the most clearly damaging aspects of the current environmental crisis—not to speak of the more complicated issues of equity involved—suggests the inadequacy of these institutions for purposes of achieving ecological sustainability.

While it is unlikely that states and markets will disappear as social coordinating mechanisms, it is imperative to empower *other* institutions more capable of managing the transition to a more sustainable world, if only at the local scale in many regions. This is what the ecosovereignty principle is designed to initiate: legitimating the power of institutions that can take the necessary measures to preserve and maintain as a "commonwealth" the natural environments that will enable peoples to sustain themselves (Purdy 2019, 97). But what will "the transition" involve, and how can the governance institutions that ecosovereignty legitimates contribute to it? While any answer will inevitably be speculative, two aspects of transition can be highlighted here.

First, as with many (though not all) aspects of environmental philosophy, ecosovereignty legitimates greater decentralization of governance. Second, because of the emphasis on ecosystem-based governance, it also advocates a transition to a more agrarian socioecological system. Some comments about both of these aspects are warranted, by way of conclusion.

The debate about decentralization, for purposes of ecological sustainability and otherwise, has many dimensions, some of them already discussed in previous chapters. Certainly, there are negative aspects of decentralized—that is to say, local, place-based—solutions. Coordination between localities, especially in terms of policies aiming to reduce environmental degradation, can be difficult. At the same time, the main arguments for local, ecoregional governance, are twofold. For one thing, such governance has greater and easier access to local knowledge about environmental conditions, and is able to mobilize this knowledge in the service of achievable goals without being subject to other interests and demands from outside the region. While there may be local political elites that could seek to exploit resources for their own benefit, ecosystem peoples have at least equal chances of resisting such exploitation without the intervention of more powerful market or state forces from outside the region. Governance structures are always capable of corruption; but power works at different scales, as does its deformation to serve elite interests.

Another argument for local, ecoregional governance points to the manifest inability and/or unwillingness of larger states and governance mechanisms to eschew goals of growth and development for those of sufficiency and sustainability. Reasons have been given earlier for why this is the case; most often, they involve the advantages to be gained in greater wealth and power from intensive resource extractions that lead to increased industrialization. What is the counterweight to these advantages, other than the organized opposition of ecosystem peoples at a federal or constitutional level to extractivism? But such opposition requires previous organization at a different, more local scale.

The idea of transition to a more agrarian society may seem less likely than the decentralization of governance. Yet, consider the disadvantages of continued urbanization for dealing with the environmental challenges of this century: the increasing expense of the fossil-fuel energy that powers urban centers, the decreasing availability of affordable food and water for the megacities, the increasing vulnerability of such urban areas to diseases (including future pandemics), the decreasing manageability of large-scale urban regions in terms of public order and service delivery, and perhaps above all, the greater scarcity of employment in automated industrial economies (Orr 1994, 622–23).

A new agrarianism involves, in the nature of the case, lower energy use, greater access to cheap food and water, better living conditions, safer environments, and greater employment opportunities. Agrarian societies are those in which people grow their own food, and use their own labor to satisfy many of their ongoing needs. In industrial societies increasingly short on affordable basic necessities for all, as well as of employment capable of providing the needed incomes for a decent life, there will be little hope of escape from the miseries of a radically diminished existence for the billions condemned to live there. Empowering ecosystem peoples to maintain their own environments could, at the same time, enable urban populations to discover a new life through an increasing demand for agrarian labor. Nonindustrial farming methods—agroecology, in contemporary parlance—will require lesser energy inputs, but greater labor inputs, in order to scale up to satisfy increasing demands for food. Nothing in urban-industrial societies can match this demand, except for poorly paid service work or unskilled manual labor. Given increasing costs for necessities, there is every reason to expect greater "social unrest" and even political upheaval in the face of this mismatch between needs and incomes.

Breaking out of this cycle of impoverishment and deprivation will require an exit strategy from industrial capitalism. Certainly, part of this will include transition from industrial to agrarian forms of life. Of course, industrial society and urban centers will continue to exist. But industrial capitalism has shown itself incapable of satisfying the needs of the billions of people that are now living on this planet. The system will shrink to a size at which it may be capable of surviving, by continuing to serve a smaller elite, based on the exploitation of the labor of a larger, but still diminished, workforce. The ecosovereignty principle serves those seeking a "conceptual tool" to understand how to break out of this system, and initiate a different, more ecologically sustainable society. That this transition is possible, and certainly desirable, is one of the conclusions to be drawn from this book.

References

Battistoni, Alyssa. 2025. *Free Gifts: Capitalism and the Politics of Nature.* Princeton, NJ: Princeton University Press (forthcoming).

Buchanan, Allen. 1991. *Secession: The Morality of Political Divorce from Fort Sumter to Lithuania and Quebec.* Boulder, CO: Westview Press.

Cafaro, Philip, and Eileen Crist. 2012. *Life on the Brink: Environmentalists Confront Overpopulation.* Atlanta: University of Georgia Press.

Crary, Jonathan. 2022. *Scorched Earth: Beyond the Digital Age to a Post-Capitalist World.* London: Verso.

Crist, Eileen. 2019. *Abundant Earth: Toward an Ecological Civilization.* Chicago, IL: University of Chicago Press.

Dahbour, Omar. 2013. *Self-Determination Without Nationalism: A Theory of Postnational Sovereignty.* Philadelphia, PA: Temple University Press.

Dann, Christine. 2023. "Earth Uprisings—une histoire." Radicalecologicaldemocracy.org

Davis, Mike. 2005. *Planet of Slums.* London: Verso.

Gardner, Steven. 2013. *A Perfect Moral Storm: The Ethical Tragedy of Climate Change.* Oxford: Oxford University Press.

Gies, Erica. 2022. *Water Always Wins: Thriving in an Age of Drought and Deluge.* Chicago, IL: University of Chicago Press.

Jamieson, Dale. 2014. *Reason in a Dark Time: Why the Struggle Against Climate Change Failed—And What It Means for Our Future.* Oxford: Oxford University Press.

MacKinnon, J. B. 2013. *The Once and Future World: Nature as It Was, as It Is, as It Could Be.* Boston, MA: Houghton Mifflin Harcourt.

Nairn, Tom. 1981. *The Break-Up of Britain: Crisis and Neo-Nationalism.* 2nd ed. London: Verso.

Orr, David. 1994. "The Effective Shape of Our Future." *Conservation Biology* 8 (3): 622–24.

Pimentel, David, and Michael Burgess. 2013. "Soil Erosion Threatens Food Production." *Agriculture* 3: 443–63.

Purdy, Jedediah. 2019. *This Land Is Our Land: The Struggle for a New Commonwealth.* Princeton, NJ: Princeton University Press.

Ribo, Ignasi. 2012. *Habitat: The Ecopolitical Nation.* London: Mycelia Books.

Vince, Gaia. 2022. *Nomad Century: How to Survive the Climate Upheaval.* London: Allen Lane.

Wilson, E. O. 2016. *Half-Earth: Our Planet's Fight for Life.* New York: Liveright Publishing Corp.

INDEX

FSC
www.fsc.org
MIX
Papier | Fördert
gute Waldnutzung
FSC® C083411

Kolibri 360 GmbH
Ferdinand-Jühlke-Straße 7
99095 Erfurt, Deutschland
produktsicherheit@kolibri360.de